Developing Technology Managers in the Pacific Rim

Comparative Strategies

E. Yegin Chen
Jong-Tsong Chiang
Charles H. Davis
Dong Guilan
Ji-Soo Kim
Cornelis P.F. Luhulima
M. Rahimah Said
Richard Smith
Chatri Sripaipan
Sun Mengxin
Wu Yikang

editors
Karen Minden and
Wong Poh-Kam

An East Gate Book

Routledge
Taylor & Francis Group
LONDON AND NEW YORK

An East Gate Book

First published 1996 by M.E. Sharpe

Published 2015 by Routledge
4 Park Square, Milton Park, Abingdon, Oxon OX14 4RN
605 Third Avenue, New York, NY 10017

Routledge is an imprint of the Taylor & Francis Group, an informa business

Library of Congress Cataloging-in-Publication Data

Developing technology managers in the Pacific Rim : comparative strategies / Karen Minden and Wong Poh-Kam, editors.
p. cm.
"An East gate book."
Includes bibliographical references and index.
ISBN 1-56324-618-X (hc : acid-free paper).
ISBN 1-56324-619-8 (pbk : acid-free paper).
1. Technology—Pacific Area—Management.
2. Technology and state—Pacific Area.
I. Minden, Karen. II. Wong, Poh-Kam.
T30.A1D48 1995
338.9'5106—dc20 95-15666
CIP

ISBN 13: 978-1-56324-619-7 (pbk)
ISBN 13: 978-1-56324-618-0 (hbk)

Contents

The Editors and Contributors

Karen Minden is vice president, Research and Analysis, at the Asia Pacific Foundation of Canada. She was previously director of policy studies and management programs, Centre for Higher Education Research and Development at the University of Manitoba in Winnipeg, Canada. She was a visiting fellow at the National University of Singapore's Centre for Management of Technology and the Institute of Southeast Asian Studies in 1992–93. Dr. Minden received her B.A. from York University in Toronto, M.A. in Asian studies from the University of California at Berkeley, and Ph.D. in political science from York University. She is also a graduate of the Beijing Language Institute.

Wong Poh-Kam is senior lecturer at the Faculty of Business Administration, National University of Singapore. He is also the director of the Centre for Management of Technology and the deputy director of the School of Postgraduate Management Studies of the faculty. He obtained his B.S. in physics and electrical engineering, his M.S. in electrical engineering and computer science, and a Ph.D. in planning from the Massachusetts Institute of Technology.

E. Yegin Chen is a partner with a venture capital firm in Hong Kong. He was a lecturer in the Faculty of Business Administration at the Chinese University of Hong Kong, where he taught business economics. He received his Ph.D. from the Massachusetts Institute of Technology Sloan School of Management. His research interests include technology development in Asia.

Jong-Tsong Chiang is an associate professor in the College of Management at National Taiwan University in Taipei, Taiwan. He has worked in industry, the Industrial Technology Research Institute, and the Science and Technology Advisory Group for the premier of Taiwan for ten years and did research on technology transfer in Japan and on technology innovation in Europe for two years. He received his Ph.D. from the Massachusetts Institute of Technology.

Charles H. Davis undertook his undergraduate education in biology in the United States, and his doctoral work in science and technology policy in Canada. He worked in the S&T policy field for ten years in Canadian advisory agencies, first with the Conseil de la science et de la technologie du Québec, and then with the Science Council of Canada. He is presently a senior program specialist with the International Development Research Centre (IDRC), a Canadian agency that supports research on sustainable development in developing countries. He is based in the IDRC Regional Office for Latin America and the Caribbean in Montevideo, Uruguay.

Dong Guilan is director of the S&T Policy Research Division of the State Science and Technology Commission's National Research Center for S&T for Development. She graduated from the Beijing Institute of Foreign Languages. In the early 1980s, after serving as program officer in the International Department of the State Science and Technology Commission for many years, she moved on to the National Research Center for S&T for Development, where she carried out policy research. In the late 1980s she spent one year at the University of Maryland as a senior visiting scholar, exchanging experience in business incubator development and human resource development.

Professor Dong's major fields of policy study are technology transfer, human resource development, and business incubators. She was the first person to introduce the concept of business incubators in China, and she made policy recommendations to government for the development of such ventures. She has published dozens of articles and research papers, translated three books, and is the author of *Business Incubator—A Bridge Between Knowledge, Capital and Market.*

Ji-Soo Kim is with the Department of Management and Policy at the Korea Advanced Institute of Science and Technology (KAIST) in Seoul. His research interests include technical human resource development, development economics in science and technology, and technical entrepreneurship. He was a board member of Volunteers in Asia and a research consultant with Hewlett-Packard's intercontinental operations. He has been involved with the human resource development working group of Asia Pacific Economic Cooperation. He is also consulting adviser to the Korea Science and Engineering Foundation. Dr. Kim received his bachelor's degree in textile engineering at Seoul National University, a master's degree in industrial engineering, and a Ph.D. in engineering economy and education in the interdisciplinary program at Stanford University.

Cornelis P. F. Luhulima is currently senior research specialist at the Center for Political and Regional Studies of the Indonesian Institute of Sciences in Jakarta. His fields of specialization are European Community studies, ASEAN, and science, technology, and international relations.

M. Rahimah Said is a senior policy maker in the Malaysian Ministry of Science Technology and the Environment. She frequently represents the ministry at international meetings.

Richard Smith is assistant professor in the School of Communication in the Faculty of Applied Science, Simon Fraser University, Burnaby, British Columbia. He is also associate director, Management of Technology in the Centre for Policy Research on Science and Technology at Simon Fraser University at Harbour Centre, in Vancouver, British Columbia.

Smith was most recently an NSERC/SSHRC research associate in Management of Technological Innovation, with industrial sponsorship from Bell-Northern Research British Columbia Telephone Company. His academic training is in communication and law, and he holds a B.A. (honours) degree from Carleton University in Ottawa, and M.A. and Ph.D. degrees from Simon Fraser University in Burnaby, Canada. He is also a principal in a computer systems integration firm, active since 1987.

Chatri Sripaipan, professor of electrical engineering at Chulalongkorn University in Bangkok, was the director of the Science and Technology Development Program at the Thailand Development Research Institute (TDRI). He received his bachelor's and master's degrees in electrical engineering in Australia and his Ph.D. from the University of Hawaii. He worked for both telecommunications manufacturing and service industries in Australia before returning to Thailand. Since 1969, he has taught electronics in the Department of Electrical Engineering at Chulalongkorn University. Before joining TDRI, he served for three years as the first director of Chula Unisearch, a business-oriented consulting unit of Chulalongkorn University. His work at TDRI is in S&T policy, industrial strategies, investment opportunities, R&D systems, S&T human resources, the telecommunications industry, and information technology, among other areas.

Sun Mengxin is presently program officer of the Division of International Organizations of the State Science & Technology Commission's Department of International Cooperation in China.

Sun graduated from the Department of International Economic Law, China University of Politics and Law in July 1987. He then received intensive training in English for one year, and began teaching English at the China University of Politics and Law in September 1988. In 1991, he published his first translation, *A Dictionary of Good English*.

Sun joined the State Science and Technology Commission in 1992. He has been engaged in international S&T cooperation with different international organizations such as the Asia and Pacific Center for the Transfer of Technology (APCTT) of the Economic and Social Commission for Asia and Pacific (ESCAP), the International Center for Genetic Engineering and Biotechnology (ICGEB), and the Pacific Economic Cooperation Council (PECC). He has attended many international conferences, including the PECC S&T Task Force Workshop for Technology Transfer held in Hong Kong in November 1993. Besides serving as a government official, he also takes part in research, including the study of protection of intellectual property.

Wu Yikang is vice chairman of the China Association for International Science and Technology Cooperation. He is a member of the China National Committee for Pacific Economic Cooperation (CNCPEC), and vice president of the Science and Technology Subcommittee of CNCPEC. He is co-chairman of the Pacific Economic Cooperation Council (PECC S&T Task Force).

Wu graduated from the Department of Architecture, Nanking University, in 1952. He began his research work in the field of architecture and urban planning in the Research Institute of Civil Engineering and Architecture, Chinese Academy of Sciences (1952–56) and Chinese Academy of Building Construction (1957–60).

Wu lectures on international S&T cooperation at the Shanghai University of Technology and conducts research in the area of international S&T cooperation; and he has published a number of articles and papers.

Preface

This book is the outcome of a collaborative study among a group of research associates from nine Pacific Rim economies. Each author's work is based on a detailed concept paper and study outline, which was developed with the assistance of an international group of experts in Singapore. The authors then met in Singapore in March 1993 to discuss their individual chapters, a comparative assessment of the country studies, and the policy implications of the research. Based on the discussions at the Singapore conference and on detailed comments from the editors, the chapters were then revised by the respective authors. Each chapter thus follows a similar outline, addressing the context for developing management of technology expertise, including a brief profile of the economic structure, industrial organization, and system of science and technology innovation in the particular country. An overview of science and technology (S&T) policy includes a historical perspective on S&T development and the strategic focus of current policies. A discussion of public sector strategies, academic policies and programs, and private sector approaches to management of technology (MOT) is provided within the framework of a definition of the MOT challenge in the respective economy. Intersectoral linkages in MOT training are assessed, as well as the prevailing influence on the development of MOT. Finally, policy implications and recommendations are considered. The book includes discussions of MOT training in Canada, Singapore, Hong Kong, Taiwan, Korea, Thailand, Malaysia, Indonesia, and China.

This study grew out of the editors' involvement with the Pacific Economic Cooperation Council's (PECC) Task Force on Science and Technology. Dr. Karen Minden, the project coordinator, was responsible for the human resource development agenda of the task force and has worked closely with many of the individuals who contributed to this project. Dr. Wong Poh-Kam is the director of the Centre for Management of Technology at the National University of Singapore (NUS), and along with the Institute for Southeast Asian Studies in Singapore, hosted Dr. Minden for the year during which this project was carried out. The Management for Change program of the Canadian International Development Agency (CIDA) generously funded the project, and thanks are due to Dave Rushton and Jack Kuhns for their expert advice on the project design. Other contributors to the project include the Centre for Higher Education Research and Development at the University of Manitoba in Winnipeg, Canada, the Canadian National Committee for Pacific Economic Cooperation, the Centre for Policy Research on Science and Technology at Simon Fraser University in Vancouver, the International Development Research Centre of Canada, the Canada-ASEAN Centre in Singapore, the Canadian High Commission in Singapore, and the Asia Pacific Foundation of Canada. The intellectual and logistical support provided by the Institute for Southeast Asian Studies and the Centre for Management of Technology at NUS were invaluable; in addition, the support and friendship offered to the project coordinator by the individuals at these two institutions are deeply appreciated. The names of the people who contributed to the project design are listed in the Appendix of this book. I am very grateful to each of them and enjoyed working with such a distinguished group of practitioners and scholars.

It has been a challenge to prepare a manuscript with participants on both sides of the Pacific Ocean. We are grateful to Felix Leo, Theresa Abel, Margaret Funk, and Trish Hergesheimer for their technical expertise and patience, and to Jane Curran for her editing skill.

We hope that the "sharing of best practices" that took place during the Singapore conference will continue, and that the experiences of each of the economies under study will reinforce the goals of the PECC Science and Technology Task Force, which are to promote more equitable distribution of S&T resources, and to foster S&T capability and development in the Pacific region. This project also reflects the goals of CIDA's Management for Change Program. Through collaborative research and discussion, we hope to have promoted a better understanding of existing conditions and policies, and thus to have made some contribution to human resource development in the integration of management and technology skills in the Pacific region.

Karen Minden
Winnipeg, Canada
April, 1994

Developing Technology Managers in the Pacific Rim

1

Introduction: The Technology Management Challenge

Karen Minden and Wong Poh-Kam

The Challenge of Technology Management

The economic dynamism of the Asia-Pacific region over the last two decades has been significantly driven by technology-based enterprises. As increasing numbers of firms seek to create competitive advantage through rapid technological innovation and utilization, there has been a concomitant increase in the demand for technology management skills. This demand for technology managers poses problems at every level of economic planning and human resource development, from the firm to higher-education institutions to national governments.

One of the challenges of technology management is the need to bridge the gap between scientists and technologists on the one hand and those who run the business side of technology enterprises on the other. The same compartmentalization plagues the public sector, where government officials often lack knowledge of the technology they seek to regulate or promote, while scientists and engineers in public research institutes often lack an understanding of commercialization and business development.

Academic models of management of technology (MOT) derive predominantly from American and European business and engineering faculties. Theoretical and empirical studies of MOT concentrate on the experience of advanced industrialized economies, with a focus on the United States, Japan, and Western Europe. This book offers a different perspective on MOT, derived from economies outside the "mainstream" of MOT thought but nonetheless challenged by escalating technological development. The case studies that follow are on the economy of Canada, an advanced industrialized country; four Asian newly industrialized economies (NIEs) of Singapore, Hong Kong, Taiwan and Korea; the ASEAN economies of Thailand, Malaysia, and Indonesia; and China, with its huge, developing economy.

The need to develop human resources to meet the demand for technology managers requires strategies for education, training, and recruitment. Some countries have responded by encouraging the temporary immigration of foreign experts; others have sent professors and students to study abroad. These are just two examples of a myriad of strategies that may be adopted to develop the supply of technology managers. Longer-term planning must address the building of an infrastructure to deliver and develop both knowledge and education. Choices have to be made about research priorities, funding, and the organization of training for technology managers.

Continuous change in technological knowledge requires that institutions accommodate for changing skills requirements. Firms may respond by developing in-house training and upgrading; governments may offer upgrading programs for the technical workforce; educational institutions may have to reorganize their conceptualization of traditional disciplines. These challenges constitute the key issues in MOT policy and require responses from the private sector, academia, and government.

The ultimate objective of this book is to inform decision makers, policy makers, and educators in both the private and public sectors about the range of management of technology policies and programs in the region. Each of the chapters provides an evaluation of the goals and programs in MOT in each of the three key sectors—industry, government, and academe. The comparative approach provides data and analysis that may be used in the development of domestic policy, and in enhancing the development of programs with international dimensions. This book provides a survey of issues, problems, and best practices in the development of MOT resources across a broad spectrum of organizations in the Pacific Rim.

We begin with a discussion of the technology management challenge as it is perceived in each of the economies under study. Who defines the issues of MOT? How do they fit into the broader context of development and of science and technology policy? Each chapter then focuses on an examination of public, private, and academic strategies for developing technology managers, and on the interaction among these three sectors. Through our analysis we seek to identify the issues faced in the development of human resources for the management of technology. What strategies have been adopted to promote this development? What are the constraints on developing MOT training? What trends and patterns can be discerned in the region? How can they be explained? What policy implications emerge from both the individual case studies and an overall comparative assessment of the region?

Technological Capacity Building and the Role of MOT

> No single technological achievement yields a lasting competitive advantage. This was true of steam power; it is also true of computers. . . . The game remains a strategic one: plotting growth, integrating jobs, thinking through products, motivating people. ("Competing Beyond Technology" 1989, 3)

Economies, both national and at the firm level, must respond to the challenge to maintain, change, or grow. This is particularly true with respect to the technological base of an economy, and it requires not only human resources in science and technology, but managers with the skills needed to cope with constant technological change and competition.

Technological capacity building involves the dynamic interaction of technology, development, and trade and needs to be managed both at the macro (national) level, and at the micro (firm) level. In other words, technological development occurs in a multilevel context including the firm, the system of innovation, and the policy system. A United Nations study of transfer and development of technology refers to "techno-economic management" as an "operational tool, at both macro- and micro-levels of decision-making in government, in the enterprise sector and in scientific-technological entities." It suggests that the core of effective development is effective management of resources. Strategic planning and implementation are critical to technological capacity building. Developing the relevant human resources is central to this process and to the transfer and development of technology (UNCTAD 1990, 59).

Technology Management Skill Requirements

The above conceptualization of MOT provides some understanding of the skills required of technology managers. In general terms, the National Research Council of the United States defines MOT as the linkage of "engineering, science, and management principles to plan, develop, and implement technological capabilities to shape and accomplish the strategic and operational objectives of an organization." It further states that "effective work in the field of MOT can play a critical role in devising the strategies and imparting the skills and attitudes to [U.S.] engineers and managers that they will need in a future technology-dominated economy" (National Research Council 1987). Using this definition as a starting point, we recognize that economies at different levels of development may require a different combination of skills to manage the process of technological development.

The types of skills required are linked to the constellation of technological capabilities at both the firm level and the national level. This concept of technological capability has been much discussed and debated in the literature on development economics. A recent synthesis of the research suggests that it is useful to distinguish between technological capabilities at the individual firm level versus the technological capabilities of an economy as a whole (Lall 1992). In essence, *firm-level technological capability* (FTC) is a firm's ability to pursue various generic value-adding activities, whereas *national technological capability* (NTC) is the ability of a nation as a whole to deploy the collective capabilities of the firms that operate synergistically within the nation.

Firm-Level Technological Capability

Although there are various ways to categorize firm-level technological capabilities (Katz 1984; Dahlman, Ross-Larson, and Westphal 1987; Lall 1992), we have used a simple classification scheme that breaks down technological capability into three dimensions: product technology capability, process technology capability, and market/services capability (see Table 1.1). This scheme provides indicative characterizations of the representative levels achieved by technologically advanced firms, middle-level firms, and low-level firms respectively. The typical competitive strategies of firms at each of the three FTC

Table 1.1

A Simple Typology of Firm-Level Technological Capabilities (FTC)

Technological Capability Dimension	Level of Technological Capabilities		
	Low	Medium	High
Product technology	Little or no product technology capability beyond ability to interpret product specification and reverse engineering of simple products	Some product engineering capabilities; more sophisticated reverse engineering skills; minor product adaptation and innovation	Basic R&D; product innovation, development, and engineering; industrial design; technology out-transfer management
Process technology	Operations and maintenance of basic process technology; minor adaptation of process operations	Some ability to evaluate process technology capability; some process engineering and operations improvement; some procurement management; ability to assimilate new more sophisticated process technology	Process development and engineering; sophisticated operations management; continuous process improvement; supplier development and management; technology in-transfer management
Complementary assets (e.g., marketing, brand name, corporate culture, value-added services, etc.)	Little consumer product marketing beyond domestic market and lower-end product based on cost advantage; mainly subcontracting for others on consignment basis	Some product differentiation through marketing and brand name development; increasing scope of subcontracting to offer turnkey original equipment manufacturer (OEM) services	Industrial packaging; own distribution channels and/or brand name; turnkey consultancy services to install new technology and improve operational productivity
Overall technology strategy	Reactive response; technology plays minor role in business strategy	Prospective planning; technology as an important but largely given element in business strategy	Proactive planning; technology as a key variable in business strategy

(continued)

Table 1.1 *(continued)*

Technological Capability Dimension	Level of Technological Capabilities: Low	Medium	High
Representative competitive strategy	Concentrate on low value-added products and compete largely on low labor cost	OEM manufacturing—compete on production quality; fast follower—compete on rapid imitation/adaptation	Technological leader—compete on rapid product innovation; quality leader—compete on product quality and features; broad cost leader—compete on economy of scale and scope
Overall technological role	User of relatively low-level technology; licensee of technology and purchaser of capital-embodied technology; little in-house proprietary expertise	Significant user of technology; some capacity to adapt and innovate; some proprietary in-house know-how, but mostly tacit and not easily transferred	Technological lead-user and innovator; systematized packaging of proprietary know-how; licensor of technology; established trademark and brand name

Source: Wong Poh-Kam, "Developing National Technological Capabilities in Asia: Lessons of Management." World Bank/Economic Development Institute and Centre for Management of Technology/National University of Singapore, Singapore, 30 March–3 April 1993.

levels are also illustrated, together with an indication of their characteristic role with respect to technology (Wong 1993).

As with all analytical schema, this is a simplified picture of what is in effect a continuum. Nonetheless, it does provide a useful categorization of the differences in technological levels observed in practice. At the risk of oversimplification, industrialized nations can be characterized as having a significant proportion of firms at the high FTC level for all three capability dimensions. In newly industrializing economies, firms are mostly at the middle level, although a sizable number are found at the high level for the manufacturing/process technology dimension. In most developing countries, firms are still predominantly at the low FTC level, according to our classification.

National Technological Capability (NTC)

The average level of FTC of the individual enterprises operating within an economy constitutes an important indicator of the overall technological capabilities of the economy. However, economy-wide technological capabilities are not simply the sum of thousands of individual FTCs developed in isolation. Because of externalities and interlinkages, there is likely to be synergy between individual FTCs (Lall 1992). Moreover, in addition to private enterprises, public sector institutions typically play a significant role in influencing the supply volume and quality of factor inputs, particularly those factors whose production entails significant public-good characteristics that are required by firms to realize their FTC. Finally, various aspects of public policy, from macroeconomic policies to specific sectoral policy, also have significant effects on how FTCs are actually deployed.

It is the complex interplay among individual firms with specific FTCs, the supply of resource factors, and competitive environments that jointly determine the overall NTC of a nation (Lall 1992). In essence, this is conceptually similar to Michael Porter's idea of an "interacting diamond" that determines the competitive advantage of nations (Porter 1990). In summary, the various components that together define the level of technological capability of a nation as a whole include the FTC of individual enterprises, the quantity and quality of human resources, public policy environments that enhance effective deployment of FTCs, and a synergistic relationship among these factors (see Table 1.2).

Technological Development Processes

Having identified the various components of technological capabilities at the firm and economy-wide level, the analytic framework addresses how such capabilities can be developed over time.

Development of firm-level technological capabilities can be analyzed in terms of movement from one capability dimension to the next, or a combination of both. Such development requires *directed efforts*, which can essentially be classified either as efforts to create new technological know-how through research and development (R&D) and other exploratory and creative learning efforts; or efforts to absorb or deploy technological know-how already existing somewhere else, through various imitative learning and know-how transfer processes. Although all technological development processes inherently involve a certain degree of *learning by doing* and require some extent of innova-

Table 1.2

Components of National Technological Capabilities (NTC)

- FTC of individual enterprises

- FTC-enabling resources and infrastructures

Quantity and quality of human resources and other FTC-enabling resource factors, the development of which is mainly through the public sector because of their externalities and public-good characteristics. These resource-supplying public sector institutions typically include:

- public sector educational and training institutions
- public sector S&T R&D institutions
- other supporting strategic infrastructures (legal protection of intellectual property rights, information dissemination services, etc.

- Competitive pressure and incentive to acquire and deploy FTC

Public policy environments providing incentives conducive for the effective accumulation and deployment of FTCs

- macroeconomic policy
- industrial structure and competitiveness of the key markets within which firms compete
- industrial structure and competitiveness of the key factor markets
- cost and accessibility of foreign technologies and manpower skills

- Synergistic relationships and institutional mechanisms

Geographic proximity, and institutional mechanisms that create and facilitate synergistic relationships among firms

- geographic proximity of interindustry linkages and clustering of industries with their suppliers, customers, and product/process technology developers and partners that facilitate information flow/technology diffusion, promote common outlook, cultural affinity, and trust, and reduce transaction costs
- institutional arrangements that facilitate resource sharing, promote cooperation/coordination of technology efforts and lower transactions costs involved in technology investments and diffusion

Source: Wong Poh-Kam. "Developing National Technological Capability for Industrial Technological Capabilities in Asia: Lessons in Management." World Bank/Economic Development Institute and Centre for Management of Technology/National University of Singapore, Singapore, 30 March–3 April 1993.

tiveness, the former often represents a more risky and difficult undertaking. Typically, though not always, the acquisition of product technology capability requires a higher effort in technology creation than is the case for process technology.

Within this technology creation-absorption spectrum, a variety of activities can be pursued. For example, creation can occur through R&D activities in private firms, universities, or government laboratories, while absorption can be achieved through internalized transfer from parent multinational corporations (MNCs) to their subsidiaries, new technology adoption and imitation by private firms, interfirm movement of staff, and students who return from overseas.

Differences in National Technological Capability Development Strategies

For developing and newly industrializing economies, the major sources for technology absorption are likely to be foreign rather than local, whereas for the large, more advanced industrialized nations, domestic diffusion and transfer is typically more important. However, countries at the same level of industrial development may emphasize different means of international technology acquisition. Similarly, the pattern of domestic diffusion of technology may also differ considerably among nations at the same level of economic development.

In general, the extent to which a nation can effectively exploit technology as a source of competitive advantage in the marketplace depends not only on the overall level of FTCs of their national firms, but also on the relative mix of their efforts in technology creation versus absorption along each of the three capability dimensions. The relative effectiveness of such efforts and their linkages are also significant. Thus, while technology creation and the selling of know-how is an important source of economic power, the ability to exploit existing technology may be even more important, as the rise of Japan as an economic power in the 1960s and 1970s illustrates. Moreover, too early an emphasis on technology creation may be socially inefficient, if the infrastructure for diffusion and the capacity for absorption and commercialization is not well developed.

This implies that countries at different stages of industrial development will need to emphasize a different mix of technology development efforts. In particular, relatively new industrializing countries that are confronting other more industrialized nations in the marketplace will typically need to pursue a "follower" strategy that emphasizes building up absorption capacity—the effectiveness of learning, assimilating, and possibly improving available technology created by others to gain competitive advantage. The public sector may need to play a significant role in this absorption-capacity–building stage. As the country moves up the technological ladder over time, the emphasis can then shift toward creating new technological know-how.

A follower strategy need not preclude a "niche" strategy, which focuses resources on technology creation in a few select areas. These areas may be relatively neglected even by more advanced countries. In some cases unique circumstances imply that technological leadership in those niche areas may confer disproportionately significant advantages.

The above framework provides the conceptual approach relayed to each of the authors who contributed to this volume. The purpose was to provide a common understanding of national technological capability, and of the component skills of MOT at various levels of economic and technological development. With this framework as a starting point, the authors prepared an analysis of the issues and trends in technology management training in their respective economies.

Key Players in Developing Technology Managers: A Tripartite Responsibility

The Role of Public Policy

Implicit in our analysis is the notion that a society has a certain capacity to acquire, adapt, diffuse, absorb, develop, and create technology for economic growth. This *social absorption capacity* is comprised of the interaction of social institutions and government policies in matching technological learning capacity and supply with utilization (Wong and Ng 1992, 3). It represents a broad view of the key elements contributing to technical progress.

An underlying assumption of this book is that public policy plays an important role in strategic planning and leadership. The country studies evaluate public resources and strategies in the context of technology management, including its acquisition, creation, and absorption. Some of the issues to be addressed in examining this are: the extent to which government takes the responsibility to develop MOT skills; the role of expatriates in supplying expertise; who identifies the required skills for MOT; and other policies that affect decisions about developing technology management resources, such as legislation stipulating employment of indigenous staff, utilization of local supplies, and specific locales of production. In other words, what is government's role, and how do MOT policies fit into the larger policy matrix?

The Role of Academia

It is assumed that higher education institutions have the responsibility for providing the infrastructure for long-term development of human resources. Examined here are the appropriate role of colleges and universities in MOT research; the curricular development needed to keep up with rapid technological change; who the consumers of MOT education are; and recruitment of MOT teachers, whether foreign experts, "returned" students, engineers, management professors, or some combination. These issues concern the perceived mandate of the institutions, and their access to resources.

The Role of the Private Sector

The private sector is on the front line regarding the challenge to develop skilled technology managers. Firms are concerned with on-the-job training, continual upgrading of skills, and recruitment of technology managers. Among the issues assessed by the country studies are the demands for technology managers that business creates in the labor market; the interaction of business with educators; the role of multinational companies in diffusing MOT knowledge internationally; and the response of small and medium-size enterprises with limited resources to provide in-house training. In cases where academia and government do not take a proactive role, their responses are driven by the demands of the private sector. In other cases, government or academia may supply expertise where there is not a corresponding market demand.

Regional Comparisons: Approaches to Developing Technology Managers

The concluding chapter of this book examines the patterns that can be discerned in approaches to developing technology managers in the Pacific region. The results of this cooperative effort are directed toward policy makers in government and in academia and decision makers in industry. The book has three objectives: to make readers more thoughtful about the conceptualization of MOT, to provide comparative data for improved policies and programs; and to share best practices across a wide spectrum of specialties and regions. Since MOT is a developing academic field, there is a dearth of literature that can be used for reference and teaching, and of material that reflects the experiences of Pacific Rim economies other than those of the United States and Japan. This is particularly true in the area of S&T policy studies with regard to enhancing national capabilities in MOT. It is important that decision makers in government, academia, and industry, as well as students in the field, have a broad understanding of MOT in a historical, political, social, economic, and comparative international context.

A comparative approach to the Pacific Rim acknowledges the existing and increasing cooperation in the region. There are numerous bilateral and multilateral arrangements in the three sectors in this analysis, including science and technology agreements between governments, industrial strategic alliances, joint ventures, multinational corporations, university exchanges, and intersectoral strategic partnerships. It is therefore appropriate to examine regional trends and developments in this context.

References

"Competing Beyond Technology." 1989. Special section, *Harvard Business Review* (November–December): 93–120.

Dahlman, C. J., B. Ross-Larson, and L. E. Westphal, 1987. "Managing Technological Development: Lessons from the Newly Industrializing Countries." *World Development* 15(6): 759–75.

Katz, J. 1984. "Domestic Technological Innovation and Dynamic Comparative Advantage: Further Reflection on a Comparative Case Study Program." *Journal of Development Economics* 16(1): 13–38.

Lall, S. 1992. "Technological Capabilities and Industrialization." *World Development* 20(2): 165–86.

National Research Council. 1987. *Management of Technology: The Hidden Competitive Advantage*. Washington, DC: National Academy Press.

Porter, M. 1990. *Competitive Advantage of Nations*. London: Macmillan.

O'Connor, David C. 1994. "Technology and Industrial Development in the Asian NIEs: Past Performance and Future Prospects." In *The Emerging Technological Trajectory of the Pacific Rim*, ed. Denis Simon. Armonk, NY: M. E. Sharpe.

UNCTAD. 1990. *Transfer and Development of Technology in Developing Countries: A Compendium of Policy Issues*. New York: United Nations.

Wong Poh-Kam. 1993. "Development of National Technological Capabilities in Singapore: Past Strategy, Future Management Challenges." In *Integrating Technology and Management: A Tripartite Approach*, ed. Karen Minden, Wong Poh-Kam, and Dipo Alam. Singapore: PECC Science and Technology Task Force.

Wong Poh-Kam and Ng Chee Yuen, eds. 1992. *Human Resource Development and Utilization in the Asia Pacific: A Social Absorption Capacity Approach*. Singapore: ISEAS.

2

Management of Technology and Technological Change in Canada: Learning and Teaching Innovation and Competitiveness

Charles H. Davis and Richard Smith

Background

Management of technology (MOT) is emerging as a field of practice, scholarship, teaching, and policy attention at a time of unprecedented concern about Canada's economic future. While Canada's overall economic growth throughout the 1980s was among the highest of any of the G-7 countries, the economy exhibited persistent problems of low-productivity growth and sluggishness in channeling investment into high value-added activities. The recession in the early 1990s hit Canada hard. In 1991, the Canadian economy contracted by 1.7 percent. This poor economic performance raised the already high levels of unemployment in the country and increased the already high level of stress on the provincial and federal governments, exacerbating the ongoing crisis of the Canadian federal political system. By the beginning of the 1990s, the net public debt/GDP (gross domestic product) ratio had climbed to nearly 50 percent—up from 10 percent only ten years earlier.

How should competitiveness be sustained? Policy makers are bombarded with advice on this issue from industry spokespersons, consultative bodies, and management gurus. Often the central message is that Canada has to get serious about moving up the technology ladder. The prime minister's National Advisory Board on Science and Technology informed him in 1991 that

> Canadians will not succeed in meeting international competition, and will therefore face a declining relative standard of living, unless we become much more adept in applying science-based technology to create a continuous flow of innovation and productivity growth. There is no more serious challenge facing Canada today. (NABST 1991a, 2)

The policy question is how to create and sustain competitive advantage through innovation in a small, wide-open economy. The issues on the competitiveness agenda range from human resources, science, finance, and policy processes to cultural values and rules of international trade (SCC 1992). Another issue, the one we address here, concerns the kinds of skills and attitudes needed to manage technology throughout the national system of innovation. Most obviously, firms require improved innovation management skills, but complementary skills are also required in the variety of public and private institutions that support technical change, in private investment institutions, and in the collective arrangements of innovation system actors such as industrial, scientific, and technical associations. Adding complexity to the issue of how to engineer comparative advantage is the growing interpenetration of the competitiveness and the environmental sustainability agendas.

When MOT was first identified as a Canadian S&T policy issue two decades ago, it was viewed primarily as an R&D management problem to be addressed through formal training and a university-based research program. MOT was thereafter largely ignored in Canadian S&T policy debates until the late 1980s.

It is curious that management deficiencies were among the last factors to be considered in the Canadian competitiveness debate. Researchers, universities, workers, bureaucrats, the political system, government labs, the tax system, public attitudes, banks, foreign firms, small firms, the resource sector, history, and teachers were all fingered as the weak links in the national system of innovation before the skills, attitudes, and competence of managers became an issue. In the late 1980s, however, the tables turned. In a 1988 report on globalization and competitiveness, the Science Council of Canada was one of the first to put the competence of managers at the center of the innovation policy agenda:

> Improving the scope of governmental support will not in itself be much help . . . when Canadian managers place so little emphasis on technology and innovation. Canada's most immediate S&T problems are the inability of many of its managers to develop and apply technology to make a profit, the low rank within managerial hierarchies for those with technological expertise, and relatively poor rewards for these people. . . . Few Canadian companies integrate technology into their strategy formulation process. Too frequently major technological choices are treated as tactical rather than strategic decisions. Or they are viewed largely in isolation, as the concern of the R&D department. . . . The proper exploitation of technology must move to the top of the agendas of Canada's directors and senior executives. (SCC 1988, 11, 12)

What is the MOT agenda, and who sets it? In Canada, the MOT agenda is multiple, and diverse actors are tackling the management of technical change from a wide variety of disciplinary and institutional standpoints. Generally speaking, however, MOT promises to provide the kind of competence that Canada requires to develop a more productive, higher value-added, sustainable economy. We will here identify and explore issues in the development of technology-management capability in Canada in the university, private, and public sectors and place technology management in the context of the Canadian debates about innovation for economic and social development. In the final section we address the question of where this might take the field of technology management in Canada.

The Canadian "National System of Innovation"

Canada is a continent-spanning federation of about 26 million people with a 1989 GDP of C$650 billion.[1] Its economy is highly differentiated by region. Manufacturing is concentrated in Ontario and Quebec; the prairie's economy is based on agriculture; fisheries, mining, and forestry dominate the north and the coasts. Canada is also a significant energy producer.

Approximately 30 percent (1989) of Canada's GDP is generated through international trade; of the G-7 countries, only Germany has a GDP to which international trade contributes a greater proportion. Canada is highly integrated into the North American continental economy. The United States and Canada have the largest bilateral trading relationship in the world, and about three-quarters of Canadian foreign trade is with the United States. Much of this is intrafirm trade. Canada also has one of the highest levels of foreign ownership of industrial assets of any advanced country. Rates of foreign ownership are about 45 percent in manufacturing overall, ranging up to 75 percent in transportation equipment, 72 percent in chemicals and chemical products, 67 percent in petroleum and coal, and 90 percent in rubber products (Porter 1991, 15). Canada is a relatively decentralized federation, and significant barriers to interprovincial trade exist. The country is organized politically along east-west lines, but economically along north-south lines. This accentuates sensitivity to issues of wealth redistribution within the federal system, while making the definition of national economic strategy relatively difficult (Leslie 1987). As one observer has commented on Canada's peculiar political-economic structure:

> Canada is a bundle of paradoxes. Its economic development has been uneven because of external and internal factors. It is both underdeveloped and overdeveloped, a resource hinterland and an advanced manufacturer, capital rich and capital poor. (Clement 1977, 7)

Overall, Canada makes a comparatively modest effort to upgrade its technological and scientific capability. Canada's gross expenditures on research and development/gross domestic product (GERD/GDP) ratio has hovered around 1.35–1.4 percent since the late 1970s. In 1989, Canada spent about C$8.3 billion on R&D. Of this, the business sector funded about 42 percent and performed about 56 percent. Industrial R&D spending is concentrated in a few very large firms. In 1987, firms spent C$4.1 billion on R&D; 37 percent was concentrated in the top ten performers, and half in the top twenty-five performers (ISTC 1990a, 24). The higher-education sector funded about 9 percent and performed about 23 percent of Canada's R&D, while the federal government funded about 30 percent and performed about 17 percent. Provincial governments, foreign actors, and private not-for-profit organizations are other, smaller components in the Canadian innovation system.

In Canada, the federal government has historically played the role of "systems integrator" of the national innovation system. National research capability was overwhelmingly based in federal agencies and organizations until the 1960s. Federal policies, programs, and S&T

institutions are the single largest component in the national system of innovation.

The most important federal outlays in support of industrial innovation are direct funding of public-sector R&D and technical services, direct funding of research in the higher-education sector via three scholarly granting councils, significant but declining support of the recurrent costs of universities, some support of industrial R&D through specialized programs, indirect support for industrial R&D through relatively generous fiscal incentives, and a variety of programs supporting labor-force training and adjustment.

The federal government spent C$5.05 billion on S&T activities in 1989–90. About 59 percent of these expenditures were intramural. In 1990 there were 290 federal scientific establishments in all regions of the country. They represented more than twenty federal departments and agencies and ranged in size from one person-year to over 4,000 person-years. Nearly 34,000 person-years were allocated to federal S&T activities in 1989–90 (ISTC 1990c). Given the prevailing environment of fiscal restraint in Canada, few expect growth in this R&D sector (Anderson and Davis 1993).

Fifty-three Canadian universities award bachelor's degrees in the sciences, and thirty-two award bachelor's degrees in engineering (Kavanaugh 1993). Canada produces about 14,000 scientists, 7,000 engineers, and 1,200 Ph.D.s a year. The federal government is the principal supplier of financial resources for university research. In 1992–93, the principal agency that supports university research, the Natural Sciences and Engineering Research Council (NSERC), spent about $C445 million on university research. In Canada, education falls under provincial jurisdiction. Federal support for recurrent costs of higher education is being phased out.

Most observers agree that Canada's overall technological effort is too modest and her R&D expenditures too heavily dependent on the public sector for her to realize her aspirations to remain an advanced economy. In 1988, the prime minister's National Advisory Board on Science and Technology compared Canada's performance on ten key indicators of S&T effort with those of major competitors (the United States, Germany, France, Sweden, the United Kingdom, the Netherlands, and Japan). On two indicators—advanced degrees awarded by population, and government-funded R&D as a percentage of GDP—Canada was in the middle. On three indicators—government-funded

R&D as a percentage of GDP, domestic patents granted per 100,000 population, and higher-education R&D as a percentage of GDP—Canada was second lowest. On five others—gross R&D expenditures as a percentage of GDP, industry-funded R&D as a percentage of GDP, international patents granted by population, scientists and engineers in the labor force by population, and number of technology-intensive industries with a positive trade balance—Canada ranked lowest (NABST 1988, 5).

Canada has developed strong technological capability in certain areas, some of which were fostered by governments as instruments of national development: railroads in the nineteenth century, and telecommunications, avionics, and some energy technologies in the twentieth century. However, in most industries Canada is a "follower" country; "Canadian technological strategy has been that of a late comer, mostly imitative and defensive" (Niosi 1991, 91). Canada, like other "small countries," is an importer of technology through purchases of machinery, services, R&D, and skilled labor (Dalpé 1988). Foreign direct investment is an important channel of technology transfer into Canada. Subsidiaries and affiliates of transnational companies use relatively current production technologies. However, manufacturers in Canada take up improved fabrication, assembly, and materials handling technologies more slowly than their American counterparts (APRO 1992).

In an avalanche of reports and studies, Canadian competitiveness has been dissected and debated.[2] Arguably the most influential and widely visible report was one commissioned from the American business strategist Michael Porter by the federal minister of industry, science and technology, and international trade, and by the Business Council on National Issues, a private lobby for the business sector. Although the Porter report covered much familiar territory, it usefully applied the famous "diamond" analysis to the Canadian system of industrial innovation.[3] Being of American origin, it also enjoyed a degree of credibility unavailable to Canadian analysts. When it emphasized to the Canadian political and business elite the special significance of science, technology, education, and innovation for international competitiveness, many Canadian observers were pleased.

The Porter report, like many before it, argues that Canada is at a crossroads in the new competitive environment (Porter 1991). Five trends are described as being indicative of underlying weaknesses:

1. Low productivity growth. Since the late 1970s, "total factor productivity growth—i.e., output growth unexplained by additional labor and capital inputs—has been almost zero" (OECD 1992, 51).
2. Relatively high unit labor costs.
3. Persistent high unemployment.
4. Lagging investment in upgrading skills and technology.
5. A macroeconomic climate that does not encourage productive investments (Porter 1991).

The report observes that Canadian exports are concentrated in "clusters" of distinct industries. With the exception of some semiconductor and computer firms, the exported upstream industries are in natural resources: materials and metals, forest products, and petroleum and chemicals. Among so-called industrial and supporting functions industries, which tend to be technology-intensive, Canada's exports are concentrated in transportation industries, mainly automobiles and avionics. Canada's automotive industry is the result of a 1966 production-sharing agreement with the United States. Among final-consumption goods and services, Canada's exports are concentrated in the food and beverage industries (ibid. 1991). The picture is the well-known one of an economy overspecialized in exports of unprocessed or semiprocessed natural resources and failing to establish itself in a range of higher value-added economic activities (see Crane 1992, chap. 5, and Gertler 1991).

S&T Policies in Canada

We turn now to a very brief analysis of Canada's S&T policies, with a view to describing how these policies are moving to address the issues outlined above. This is not an easy task, because in addition to the federal government's policies and programs in support of industrial innovation, all provincial governments have policies and programs, some of which (e.g., those in Quebec, Ontario, Alberta, and British Columbia) are extensive and long-standing. Furthermore, like most other North American cities, Canadian cities and metropolitan areas are developing initiatives designed to stimulate and foster innovation. Our discussion focuses mainly on federal policies.

Of the three policy "trails" to Canadian innovation policy (i.e., science policy, trade policy, and industrial policy; see Doern 1990), the

Progressive Conservative federal government, which was in power from 1984 to 1993, was mostly concerned with trade policy and macroeconomic policy. Industrial policy was officially frowned upon. At the federal level, science policy has been a minor dossier looked after by a junior minister of state. In provincial governments, science policy is usually located within ministries of higher education. The Conservative government's medium-term economic policy, announced in 1984, had two basic objectives: to "create a stable and predictable domestic macroeconomic environment for the private sector" (OECD 1992, 32) through reduction of fiscal deficits and controlling inflation, and to "promote economic growth through the implementation of structural policies" (ibid.), mainly through trade liberalization and privatization. Neither of these economic policy objectives referred specifically to S&T, but both had implications for the kind of innovation policy regime established since 1984 in Canada.

No major S&T initiatives have been launched by the federal government since the InnovAction program was announced in 1988. Under InnovAction, which had a budget of about C$100 million per year for four years, five themes were pursued: fostering innovation and technology diffusion, developing capacity in three groups of strategic technologies (biotechnology, advanced materials, and microelectronics), improving the effectiveness of federal S&T efforts, building human resources, and promoting an "S&T culture" in Canada. In 1989, in an associated initiative, the federal government held a national competition under the Networks of Centres of Excellence program. This program provided support for interinstitutional groups of university and industry researchers, mainly in advanced technologies. Many programs, instruments, and incentives to stimulate industrial innovation currently are in place in Canada. These include one of the most generous R&D tax-incentive regimes in the world, applied research assistance delivered through national and provincial industrial extension systems, special R&D support programs in such areas as defense, artificial intelligence, and strategic technologies, and regionally specific technology development funds in some provinces.[4] In 1989–90, the federal government spent about C$2.7 billion to assist nonagricultural business, but regional development assistance accounted for 41 percent of the total, and energy and resource development assistance accounted for another 27 percent. Industrial or technology program expenditures accounted for 23 percent, or C$627 million, but of this, defense industry assistance represented fully 40 percent of expenditures. The trend is to reduce direct support to industry, to make a sharper distinction between industrial and regional development assistance, and to increase the volume of repayable assistance (Davis 1993).

Canadian policy style does not favor general indicative planning in which targets, measures, or goals are put forward. The Canadian political system lends itself primarily to de facto, implicit S&T policy making. It rarely defines explicit objectives such as market share or R&D spending targets. Canadian S&T policy relies a great deal on consultative processes and consensus building. In addition to its process orientation, Canadian S&T policy tends to concentrate on providing infrastructure, an appropriate environment, and discrete incentives to induce desired changes in behavior in industry and science. Universities and public laboratories, because they are accessible to public policy instruments and dependent on public financing, are attractive targets for attempts to create surrogate industrial research activities (NABST 1988, 9; SCC 1992, 74–76). Most Canadian policy initiatives not targeted on the university sector involve promotion of consortia and strategic alliances in advanced technologies, focusing on the activities of public laboratories (OECD 1990, 123–28), and on promotion of concerted action among innovation system players (Davis, Alexander, and MacDonald 1990). There are some indications that the federal government is considering strengthening institutions and incentives favoring technology acquisition and diffusion (NABST 1992).

In 1991 the federal government launched a Prosperity Initiative (so called because of the public's perceived resentment of issues of "competitiveness"), an attempt to create consensus in Canada about actions to promote economic and social well-being. The Prosperity Initiative was steered by a group of eminent persons from the private and higher-education sectors, and it supervised the production of many background reports and undertook extensive consultation with the public, industry groups, and government clients in dozens of round tables and workshops. A Task Force on Challenges in Science, Technology and Related Skills also conducted consultations and advocated a course of action (Prosperity 1992b). The Prosperity report puts forward fifty-four recommendations in an action plan. In general, recommendations fall into eight classes of initiatives:

- Collective action (a National Quality Institute would launch "a coordinated national effort to adopt a quality approach in all areas of Canadian society")
- Education and training (many proposals)
- Regulation of government (as in the proposals to independently review tax and fiscal policy, assess the impact of government on competitiveness, etc.)
- Administrative coordination (consolidation of government programs of assistance to industry)
- Tax and fiscal measures
- Infrastructure (a broadband electronic network, a Centre of Excellence for Sustainable Development)
- Legal measures (harmonization of laws regarding investment, etc.)
- Persuasion (a communications campaign to convince Canadians that learning is necessary, etc.)

Many of the recommendations are addressed to actors other than government, especially universities, communities, industry associations, firms, and trade unions, and few recommended initiatives are predicated on substantial increases in public spending (Prosperity 1992a).

Roobeek (1990, 230) observes that in industrial countries in the past decade, "technology policy hangs in the air instead of being firmly rooted in an overall restructuring policy." Canada is perhaps moving toward a more comprehensive restructuring policy in which S&T and innovation policy are elements. This may be something of a departure from the neoliberalism of the 1980s, which was concerned primarily with controlling inflation, interest rates, and public spending. The attempted restructuring involves experimenting with a new style of political action emphasizing mobilization of Canadians through consultation and persuasion, with modest doses of government financial commitment.[5] It represents an attempt to stimulate institutional and social change. The primary lesson it contains is that successful adjustment to marketization is not a passive process. However, the basic model of economic development upon which Canada has depended—export-led growth with a specialization in staples and some managed sectoral trade—is not being questioned.[6] Furthermore, after the minister for industry, science, and technology announced his decision to leave politics, what remained of the Prosperity Initiative was disbanded, and no serious follow-through has occurred. Plebiscitary technology policy has not been notably successful in Canada.

Environmental sustainability is another S&T policy issue of growing importance. The Canadian government has begun to increase the salience of environmental initiatives within its broader political and economic policy positions. The assumption is that there is a domain of compatibility between economic growth and environmental sustainability. The most notable aspect of Canadian environmental policy is the 1991 Green Plan, a six-year, C$6-billion set of coordinated initiatives in pursuit of environmentally sustainable development. Many initiatives undertaken within the framework of the Green Plan have an S&T dimension. Among those are a C$50-million fund for training the next generation of environmental scientists; a C$4.5-million Canadian Global Change Program; and the Environmental Innovation Program, which makes C$1.7 million available in 1992–93 to support extramural R&D to attain Green Plan goals.

The conclusion of a trade agreement with the United States in 1988 and the signature of the North American Free Trade Agreement (NAFTA) in 1993 with the U.S. and Mexico were the culmination of two decades of debates between "nationalists," who argued that the goal should be to foster Canadian-controlled firms able to use technology competitively, and "continentalists" who favored enhancing market mechanisms to drive Canadian industrial productivity (see Blais 1986). Doern notes:

> The free-trade initiative was built . . . on the proposition that Canada's manufacturing sector faced a serious productivity gap that could only be addressed by the cold shower of being exposed to full competition and by obtaining more secure access to its biggest market, the United States. (Doern 1990, 56)

The Canada–United States agreement limits signatories' latitude to implement policies of economic nationalism which, in Canada, are "essentially, policies supporting indigenous manufacturing industry at the expense of resource industries, and [policies] supporting Canadian capital against foreign capital" (Leslie 1987, 13). In addition to establishing trade rules for standards, intellectual property, rules of origin, and dispute settlement, the agreement restricts the signatories' latitude to screen investments and impose performance requirements such as

R&D spending targets or local procurement measures (Davis 1993). No agreement was reached regarding the definition of countervailable subsidies. The federal government is moving away from spending money on industrial development subsidies. Increasingly, Canadian S&T policies for industrial innovation will focus on the research, education, and training end of the innovation spectrum on the one hand, and on creating an appropriate infrastructure and institutional and fiscal environment for innovation on the other (ibid.).

This is the policy context in which the issue of MOT change is being discussed in Canada. MOT change within the firm is viewed as part of the larger challenge of social innovation. All social and economic actors are being asked to innovate. Improved management skills presumably are required throughout the national system of innovation, including enhanced skills and deepened understanding of industrial innovation among those who are not labeled "technology managers."

"Innovation management" therefore overlaps MOT, which has developed around a research and teaching agenda that is primarily firm-centered and generally focused on problems confronting individuals in technical environments. MOT has undergone a surge of growth in universities and to a lesser degree in the public and private sectors. In the following section we describe emerging patterns of MOT activities in Canada. Later, we examine the arguments and proposals for the upgrading of management skills put forward in the competitiveness debate in Canada.

Definition of the MOT Challenge

A thorough review of research on managing technological change in Canada was prepared in the late 1980s under the auspices of the Social Science Federation of Canada with the financial support of the Social Sciences and Humanities Research Council (SSHRC). The federation's report adopted a quite broad view of MOT to include S&T policy, management of innovation, innovation and the labor process, industrial policies and strategies, the geography of innovation, standards and standards writing, technology assessment, and risk management (Salter and Wolfe 1990). From this review a list of current research issues and topics on management of technical change was drawn up (see Table 2.1) and four "paradigms" of MOT research (see Salter 1990, 18–22) identified.

Table 2.1

Current Social Science Topics and Issues in MOT in Canada

- The stimulation of innovation and R&D within the firm
- The management of the innovation process
- Technology transfer within Canadian industry or from parent or foreign firms
- Technology transfer for export
- Technology transfer and overseas aid/development
- Organizational adoption of and adaptations to new technology
- Job classifications and job skills (related to technological change)
- The impact of technology upon the workforce in general and in specific industries
- Training issues and policies
- S&T education
- Public education concerning S&T
- Labor–management relations and collective bargaining (as affected by technological change)
- S&T
- Industrial strategy
- Trade polilcies
- Trade liberalization and its effects on the technological behavior of firms
- Standards
- Intellectual property issues
- Regulatory issues
- Technology assessment and risk analysis
- Office automation studies
- Gender and technology
- Spatial aspects of technology
- Regional and community development
- Decision theory and decision support systems

Source: Adapted from Salter (1990, 38).

According to the federation's report, the "theoretical" paradigm is "perhaps the best developed" of the four paradigms of MOT research. Scholars from all social science disciplines participate in "the study of the relationship of technology to society." This paradigm is populated by researchers who view themselves as social critics and philosophers rather than social scientists.

The "labor process" paradigm "originated in the broader literature concerning the transformation of the labor process under the conditions of modern industrial production." Originally largely Marxian in orientation, research in this paradigm is taking a utilitarian turn as methods are developed to facilitate the introduction of new technologies in the workplace and to mitigate their adverse impact.

Within the "social science" paradigm, researchers produce work along disciplinary lines primarily for scholarly (rather than practical) use. The research is largely empirical, nonevaluative, and noncommittal with respect to the value of technology.

The "business paradigm" is described by a frequently cited definition of the MOT field: MOT "links engineering, science, and management disciplines to plan, develop, and implement technological capabilities to shape and accomplish the strategic and operational objectives of an organization" (NRC 1987). The key premises of this paradigm are that (1) scientific and technical disciplines must join forces with management disciplines and that (2) the goal of this collaboration is to improve the performance of the organization (generally, the firm).

Most social science disciplines and many humanities disciplines are increasingly concerned with the significance and origins of technical change and its relation to social structure and cultural values. However, the business paradigm, much more than the other three, has succeeded in establishing itself as the growth point of MOT.

In the remainder of this chapter we focus mainly on MOT of the business paradigm variety, although we also have something to say about a growing overlap between S&T policy issues and MOT issues. We review recent initiatives in the university, public, and private sectors concerning teaching, training, research, and consulting activities in MOT, and we identify how issues of innovation management skills arise in the S&T policy debates and how MOT might evolve to deal with issues of environmental sustainability.

Organization of MOT in Canada

MOT first appeared as a significant issue on the Canadian S&T policy agenda in the late 1960s. In its comprehensive review of science policy, the Senate Special Committee on Science Policy (the Lamontagne Committee) observed that there was an "urgent need to improve the training of R&D managers and to do more research on the R&D and innovation processes" (Lamontagne 1973, 803). The initial Lamontagne policy report recommended that a committee comprising representatives from the Ministry of State for Science and Technology, Canadian management schools, and the Canadian Research Management Association develop a training program for R&D managers and a research program on "the organization of R&D activities and of innovation strategies." Two research centers were to be identified in Canada to carry out the research and training, and a scholarship and research-funding program was to be established (Lamontagne 1972, 529). However, subsequent consultations with industry and professional groups revealed a strong preference against the establishment of scholarships for training R&D managers in favor of continuing training for practicing R&D managers, fellowships for experienced (i.e., mid-career) managers, and collaborative training with industry (Lamontagne 1973, 803).

The research-funding component was implemented in 1974 by the Department of Industry, Trade and Commerce in the form of a Technological Innovation Studies Program. Between 1974 and 1985, when it was eliminated, this program funded 103 studies of innovation in Canada and distributed as many as 800 reports per month to interested parties (Clarke 1990). The program was also intended to award scholarships and fellowships in MOT innovation or technical entrepreneurship, but none was ever awarded. However, the research-funding component of the program played a critical role in seeding MOT research capacity in Canadian universities.

In 1979 the Canadian Advanced Technology Association, the Innovation Management Institute (now defunct), and the York Enterprise Development Centre sponsored a national workshop on technological innovation management education (Clarke et al. 1979). Included in the workshop's background material were the results of a survey of perceived educational needs and examples of university course curricula. The workshop identified the following for consideration with regard to MOT education and training:

- MOT encompasses management of technological innovation and technical entrepreneurship.
- Traditional engineering and science courses do not adequately train students for work in technology companies or for entrepreneurial activities.
- Entrepreneurs can and should be provided with MOT skills to enhance their success rate.
- University MOT courses should be made available to science students, engineering students, and business students.
- An introductory or overview MOT course should be provided and perhaps be made compulsory.

- Case material, contact with "live" entrepreneurs, and audiovisual teaching approaches should be used.

Throughout the 1980s, Canadian management and engineering schools established MOT courses and occasionally programs (see below). MOT was given a boost in visibility within the university and business communities through two national workshops on management and technology held in 1985 and 1988. These workshops brought together deans of faculties of management and engineering with corporate executives. Participants readily agreed that international competitiveness required new management skills, that the provision of these skills should be a lifelong process, that more research in MOT was necessary, that MOT implied more than managing hardware, and that providers of education and training should be as responsive as possible to industry needs. The workshops successfully disseminated information about university activities in MOT and about current views on MOT. However, the workshops did not generate impetus for cooperation among MOT actors in Canada. The second national MOT workshop produced a plethora of recommendations but no focused agenda (see MTI 1988). This lack of cooperation and focus may in part be attributable to rivalries among management schools and between management and engineering schools, which appeared to hinder concerted effort to develop MOT in Canada.

Public Sector Strategies

The Canadian public sector is involved in MOT education, research, and capacity building in a number of ways. As we show later, many of the MOT initiatives in the university sector and some in the business sector draw on public support at some stage. In addition to their efforts to stimulate MOT teaching, research, and conferencing activities in the business and university sectors, some Canadian government agencies are strengthening their own capabilities in MOT. Industry, Science and Technology Canada (ISTC, now known as Industry Canada), the flagship federal S&T agency, offered a course of six MOT training modules to staff across the country in 1990–91. The rationale was that "understanding the impact of technology management is important to the effectiveness of ISTC managers and officials, particularly those engaged in the day-to-day delivery of programs, services and counseling to corporate clients. Helping business to recognize the importance of managing technology will be an essential part of the ongoing role of ISTC sectoral and regional officials" (ISTC 1990b).

Each module was a stand-alone training package at the first-year M.B.A. level targeted at ISTC managers and officers in the regional offices, sector branches, policy directorates, and regional service bureaus, as well as general and senior managers from industry. Training for each module lasted three days. The first module provided an introduction to MOT. The other modules provided instruction in technology and business strategy, advanced technologies for manufacturing and service industries, strategic alliances and partnering, new product and process development, and international marketing.

A second issue concerns the ways that governments become directly involved in the provision of MOT services to firms. Canada has a complex public infrastructure, supplied by the federal, provincial, and local governments (sometimes in collaboration with educational or industrial organizations), to deliver technical support to firms. The two largest components are the National Research Council's Industrial Research Assistance Program (IRAP), an industrial technology support network of about 160 offices across the country, and the provincial research organizations (PROs), which offer applied research and technical support services. Many other technical support institutions have local, sectoral, or dedicated technology transfer functions. Few of these institutions combine technical consulting with services addressing managerial and organizational problems related to competitiveness.

The Canadian federal and some provincial ministries of industry, however, have programs that support part of the costs of technology diagnostics, business planning, strategic planning, or evaluation of firms' technology implementation strategy. Examples of federal services are the Manufacturing Assessment Service, the Advanced Manufacturing Technology Application Program, Interfirm Comparisons Service, sector campaigns, the Manufacturing Visits Program, Workshops on Informatics for Senior Executives, and Canadian Awards for Business Excellence (NABST 1992, A2). Typically these management-support services are delivered through consultants.

Recently, Quebec's Conseil de la science et de la technologie commissioned a comprehensive report on the current and future MOT needs of the province (Lefebvre, Lefebvre, and Le Luel 1992). The report recommends government action to support a "technology watch," promote international activities of firms, reduce product-service innovation time, and improve MOT capability in Quebec.

Academic Strategies and Programs

In the second half of the 1980s the teaching of MOT spread rapidly throughout the Canadian university community. A recent survey revealed that there are at present more than seventy courses related to MOT available at twenty-five Canadian universities (Clarke 1992). A similar survey five years ago found less than fifty courses at fewer than twenty universities (Clarke and Reavely 1987). Most of the growth in MOT courses in Canadian universities occurred between 1987 and 1990; little changed between 1990 and 1992 (CAC 1992, 26). At present about two-thirds (forty-five) of the courses are offered by about one-third (eight) of the universities. The concentration of MOT teaching activity in Canada is highest in the universities of Alberta, Calgary, Carleton, Ottawa, Toronto, and Waterloo, and the École Polytechnique de Montréal, and Université du Québec à Montréal.

In addition, the University of British Columbia has recently launched an MOT program, and Simon Fraser University, which offers an engineering extension program in technology management, has offered MOT courses in collaboration with the University of Waterloo and is developing plans to increase the number of graduate courses offered in MOT.

Some of the principal university-based groups and institutions active in MOT are described below.

The National Centre for Management Research and Development (NCMRD) was established in 1984 at the University of Western Ontario in London, Ontario. The center, which benefits from support from Industry, Science and Technology Canada (now known as Industry Canada) has two roles: to "stimulate and provide direction for management research being conducted at both the center and other institutions," and to "disseminate research findings to the management community." NCMRD carries out or sponsors research in a number of areas, including MOT and innovation. It has a discretionary budget with which to fund research.

The Centre de recherche en développement industrielle et technologique (Research Centre on Industrial and Technological Development), or CREDIT, and the *Centre de recherche en évaluation sociale des technologies* (Research Centre on Social Assessment of Technology), or CREST, were established as interdepartmental research centers by the Université du Québec à Montréal (UQAM) in 1986 with support from the Quebec Ministry of Higher Education, Science and Technology. CREDIT links researchers from various departments in UQAM and other universities (École Polytechnique, Laval, Sherbrooke, Montreal, and Carleton). Its researchers have investigated a broad spectrum of problems in S&T policy, industrial economics, and MOT. CREST links researchers from UQAM, McGill, and École Polytechnique. Its researchers have focused on the processes of technological innovation and their interaction in labor, management, and production environments; the acquisition of knowledge and skills; and the sociology of scientific and technical controversies. In 1993 UQAM amalgamated CREST and CREDIT to create CIRST, the Inter-university Centre for Research on Science and Technology.

The Research Centre for Management of New Technology (REMAT) was established at Wilfrid Laurier University in Waterloo, Ontario, in 1985. REMAT seeks to help Canadian managers acquire and implement new technology, to provide a forum for industry, government, and labor to resolve problems related to technology in the workplace, and to develop practical research programs to improve the capability of Canadian managers to deal with new and emerging technologies. REMAT hosts a monthly technology forum for senior executives, a series of seminars and workshops on a wide range of topics (supervision in a technology environment, quality management), and occasional conferences on topics such as technology and waste management. It also conducts an annual survey of technology, offers a technology auditing service, maintains a bibliographic database on implementation of change in technological environments, and conducts contract research. Wilfred Laurier University has offered a technology M.B.A. since 1986.

The most intensive initiative to strengthen MOT in Canadian universities was the establishment in 1990 of a program of Chairs in the

Management of Technological Change, funded by two of Canada's national granting councils for academic research: the Natural Sciences and Engineering Research Council (NSERC) and the Social Sciences and Humanities Research Council (SSHRC). A competitive program of MOT chairs funded under a matching-grant formula was developed jointly between SSHRC and NSERC. The granting councils match private contributions in establishing the chairs, which ensures industry support of and commitment to the chair program. The objective of the NSERC-SSHRC program is to encourage study, teaching, and training in the management and facilitation of technological change. According to the councils, "the driving force in global competitiveness is technological change"; thus the program addresses the problem of maintaining living standards and creating employment in the face of global competition. According to published reports, the five-year commitments from the two councils total C$4.61 million, or an average of C$922,000 per year (Caughey 1993, 2).

By the end of 1992, eight chair proposals had been approved in the first two rounds of competition. NSERC and SSHRC have approval for two additional rounds, for three chairs each, and the first of these is under way. The eight existing chairs are described below.

The University College of Cape Breton, Nova Scotia, has created a chair in Environmental Technology in Non-Urban, Resource-Based Industries located in the Tompkins Institute for Human Values and Technology. The industrial sponsors are the Nova Scotia Power Corporation, Sydney Steel Corporation, and the Cape Breton Development Corporation. This chair's research program will examine industrial innovation in resource industries in nonmetropolitan regions with regard to their competitiveness and environmental sustainability.

The École Polytechnique de Montréal has two chairs, both located in the Department of Industrial Engineering. The first, in Technology and International Competitiveness, is co-sponsored by Jarislowski, Fraser & Co., the Quebec government, and a consortium of consulting engineering and manufacturing firms. This chair will support research on concurrent engineering and strategy formulation. The second chair, in Choice and Management of Technological Projects, is sponsored by Bell Canada. It will examine problems related to the choice of technological orientations, the development and implementation of management methods appropriate for technology projects, and the concept of "optimum timing" in applying the results and marketing of the products generated by innovation and R&D.

The Hydro Québec Chair for Research in Management of Technology is located in the Department of Administrative Sciences at the Université du Québec à Montréal. The chair concentrates on analyzing management challenges facing Canadian industry. It will "promote reflection and dialogue between researchers and managers with a view to promoting technological innovation and stimulating entrepreneurial spirit in the high-technology sectors."

Carleton University's chair, located in the business school, is in Managing Technological Change in Manufacturing. The chair examines "the role of manufacturing in the front end of the product development cycle, the establishment of manufacturing as a core competency of a firm, and the creation and management of external linkages anchored on manufacturing for competitive advantage." The industrial sponsors are Gandalf Technologies, Lumonics, Mitel, and Northern Telecom. A Research Centre in High Technology Management established in 1984 with provincial support within the Carleton University Business School has since closed.

The NSERC/SSHRC Alcan Chair in Management and Technology is located in the business school at Queen's University. The sponsor is Alcan, Ltd. This chair was originally established in 1986 as an NSERC industrial chair and then integrated into the NSERC/SSHRC program. The chair's work focuses on management of R&D functions, technical entrepreneurship, commercialization of university-based research, and strategies for high-technology firms.

The chair at the University of Toronto, in Management of Technological Change in the Financial Services Industry, is shared by the Faculty of Management and the Faculty of Engineering. The industrial sponsors are the Royal Bank, the Toronto Dominion Bank, and Stentor Alliance (formerly Telecom Canada). This chair emphasizes research on the management of computer and communications technologies in the financial services industry. It also forms the core of a new center for the Management of Technology and Entrepreneurship.

Simon Fraser University and the University of Waterloo share a chair in Management of Technology: Innovation and Change. The industrial sponsors are Bell-Northern Research and British Columbia Telephone Company. The research program is aimed at finding ways to reduce the time it takes to formulate and successfully market technologically advanced products and services. The research looks at multidisciplinary formulation teams and their management. This chair is also interested in the use of wide-area network links and teleconferencing for teaching and research collaboration, and direct electronic links have been established with some of the MOT chairs at other universities.

In addition to the chairs program, SSHRC supports academic research on MOT under two strategic grants program themes, Science and Technology Policy and Managing for Global Competitiveness. Both program themes were established in 1989, but earlier, somewhat analogous themes (Managing the Organization in Canada and Human Context of Science and Technology) were initiated in 1983. Through these programs a total of C$2.7 million was awarded in 1992 (Caughey 1993). NSERC also supports research in the discipline of management science (C$36,000 in 1992), and research whose primary area of application is management science (C$747,000 in 1992). If we include the SSHRC scholarships in science policy (C$150,000 in 1992), the two councils currently distribute just over C$4.7 million per year in studies broadly related to MOT, innovation, entrepreneurship, and S&T policy.

This sketch of MOT initiatives in Canadian universities illustrates the growing interest in MOT teaching and research in Canada. Canadian researchers from academia, industry, and government actively participate in key North American MOT conferences. At the PICMET* conference in Portland in 1991, Canadians authored or co-authored 23 of the 278 papers presented. The Third International Conference on Management of Technology, held in Miami in 1992, also included a large contingent of Canadian MOT researchers. Canadians authored or co-authored 24 of a total of 147 papers. In each conference, the most popular theme for Canadian presenters was R&D management.

The above sketch also illustrates the importance of public and private sponsorship in establishing academic institutions supporting MOT teaching and research. Before the 1980s, MOT activities in Canadian universities were scattered throughout faculties of business or engineering. Two kinds of university-based institutions were created in the 1980s: centers and chairs. The centers usually begin with university and government support and then seek to solicit private sponsorship through membership fees or the sale of services. This movement to

---

*Portland International Conference on Managing Engineering Technology.

secure university-based MOT centers through private support has had some successes, but growth appears to be slow. The MOT chairs program, which resulted in the reinforcement of capacity very quickly, was designed from the beginning to require private sponsorship.

No university-based networks, associations, or organizations for MOT exist at the national level in Canada. A proposal for a Technology Innovation Management Network submitted to the federal government's Networks of Centres of Excellence program in 1989 was not successful. By way of comparison, in Britain, the JUPITER consortium (Joint Universities and Polytechnics Industrial Technology Education and Research) provides a network of courses offered by participating educational institutions, including the Short Course Compendium and Training Guide in Technology Management.

One pan-Canadian initiative deserves mention. This is the Technology Management Case Competition, the first such case competition in North America, launched in 1990 by the University of Saskatchewan and currently coordinated by the University of Ottawa. This competition is open to students from Canadian universities and is supported by a consortium of private and public sponsors. Each university can field one team of four members, of which at least one member must be a business student and one an engineering student. The cases "center on a major business and technology problem faced by a Canadian company." Registered teams receive the case by fax and have five hours to prepare a solution in the form of a recommended strategy. Winning teams receive honors and prizes. In the first competition, nineteen teams from twenty-one universities participated.

In April 1993, the first meeting of what may become a Canadian MOT network took place in Montreal. Attendees included the NSERC/SSHRC Chairs in the Management of Technological Change as well as numerous other researchers, research administrators, and technology managers from industry. A committee was struck to examine the best approach for Canadian participation in the International Association of Management of Technology (IAMOT).

CANMOT, the Canadian Association for the Management of Technology, was established in 1993, with a mission to support the development and diffusion of the new science of management of technology in Canada. CANMOT will be the first affiliate of IAMOT, a new international organization with a comparable mission, which held its fourth annual meeting in October 1993.

CANMOT will

- promote the advancement of technology management and innovation management by regular meetings and communications to transfer ideas on research, education, and best practices;
- provide an interactive forum between academic researchers and corporate practitioners of MOT;
- bring together scholars, practitioners, corporations, associations, and other organizations concerned with the management of technological change;
- encourage research and education in the field;
- be a source of strategic information and intelligence.

We have searched for reliable information about the market for graduates of MOT programs in Canada. Mark Abbott's 1988 study suggests that employers prefer on-site or customized MOT training, and that they do not value M.B.A. graduates, especially young ones, who are viewed as having unrealistic expectations about what they can contribute to a firm. The best candidates for MOT training, in the view of these employers, are people with strong technical backgrounds and five to eight years of work experience. Davis and Tiffin (1992) found that few executives were aware of the field of MOT. Their survey of executives' perceptions of technology management problems identified five distinct viewpoints, each reflecting an understanding of how technology contributes to a firm's competitiveness and each representing a market segment with specific requirements for MOT skills, knowledge inputs, and services. They called the viewpoints *dependent optimization*, *independent product innovation*, *global innovation*, *potential growth-driven innovators*, and *technological entrepreneurship*.

Private Sector Organization for MOT Training

Concurrent with the wave of interest in MOT in universities have come numerous initiatives to establish income-generating MOT activities in the private sector. A private market for technical management services and educational events is developing in Canada, although we have not yet located a reliable survey of this market and its suppliers.

The most important institution in Canada offering MOT services on an income-generating, not-for-profit basis was the Management of Technology and Innovation Institute (MTI). This institute was established in 1987 at McMaster University in Hamilton, Ontario, as a national center with joint sponsorship from the federal government, McMaster University, and a number of firms. In 1989 corporate sponsors were Atomic Energy of Canada, Domtar, Fell Fab, GM Canada, H.L. Blachford, IBM Canada, Lumonics, Northern Telecom, Nova, Ontario Hydro, ORTECH International, Stelco, Westinghouse Canada, and Zepf Technologies. MTI closed its doors in December 1992, heavily in debt. MTI was an independent organization "developed in response to the crisis in international competitiveness and technology exploitation faced by Canadian industry" and "dedicated to helping Canadian companies compete successfully through more effective use of their technological and human resources" (MTI 1990). MTI offered workshops, conferences, training seminars, and consulting services. The training services and seminars were offered in major cities throughout the country. In its fall 1990 program MTI offered a lineup of twelve seminars.[7] Additional on-site seminars and custom corporate services were available on demand. MTI had a small in-house staff. Its strategy was to keep overhead low by using university teaching staff from around the country in the seminars and workshops it offered. MTI also hoped to generate income through the preparation and sale of MOT audiovisual teaching material. MTI was expected to recover costs and eventually become largely self-supporting through sales of memberships and services. MTI's failure, even when benefiting from access to top university-based MOT researchers and significant corporate, government, and university sponsorship, suggests that there is a limited market in Canada for MOT training services targeted at mid-level decision makers.

A number of Canadian consulting firms large and small are offering research, educational, and project management services in MOT and S&T policy. Stargate Consultants of Ottawa conducts studies of innovation and innovation policy, performs evaluation of R&D programs, and offers management training programs in R&D management, technological innovation, and technical entrepreneurship. Pallister-Tiffin Technology Management, recently closed, was a dedicated MOT management consulting firm based in Calgary and specializing in advanced technology firms and firms in the resource sector. Terracy, in Vancouver, provides an international management and consulting service in advanced technology, venture management, and business development, with a focus on creative innovation and environmental efficiency. Hickling, with offices in Ottawa, Toronto, and Washington, offers a wide range of services in technical management and S&T policy, including decision analysis, human resources management, technology assessment, environmental technologies, and management of technology. Policy Research International and Cognetics, based in Ottawa and Vancouver, respectively, are active in S&T policy and MOT projects in developing countries. Nordicity/NGL, Doyletech, and RockCliffe Research and Technology are three well-known Ottawa firms that offer services in S&T policy analysis, economic development, project management, and investment management. Other consulting firms known to have interests and activities in MOT and S&T policy are Prospectus, Policy Impact Associates, Clarkson Gordon, Arthur D. Little Canada, Ernst and Young, Peat Marwick, the Canada Consulting Group, and SECOR.

The national profile of private suppliers of MOT services has never been established. The Quebec Council of Science and Technology examined the availability of private technology management services in the province. It found that of 1,059 management service firms in operation, 95 percent were in informatics and office automation. Less than 4 percent offered services in the domain of industrial planning (CSTQ 1990, 111).

Only one private think tank in Canada has developed a major stream of MOT activities and services. This is the Conference Board of Canada, a not-for-profit organization established in 1954 and affiliated with similar organizations in the United States and Europe. The board's mission is to be "the leading independent applied research institution dedicated to enhancing the performance of Canadian organizations within the global economy." The Conference Board of Canada has more than seven hundred associates (member firms and organizations) from across the business spectrum, and it provides access to more than three thousand associates through its American and European affiliates. It publishes the *Canadian Business Review* and pitches its services at business executives. It conducts research and undertakes economic forecasting and analysis on public policy, certain business sectors, organizational effectiveness, and international business issues. The board communicates the results via conferences, speakers programs, round tables, newsletters and executive summaries, bulletins, and reports. Since 1983 the Conference Board of Canada has had a Management of Innovation and Technology program with the following components and activities:

- the Council for the Management of Innovation and Technology, a forum for senior corporate executives from about forty companies to discuss strategic issues concerning technology management;
- a Research Managers Forum, which focuses on the needs of intermediate research managers with day-to-day R&D management responsibilities (approximately fifteen companies are currently members);
- an annual conference on the Management of Innovation and Technology, which focuses on "key factors affecting the successful integration of innovation and technology in business operations";
- occasional seminars and workshops, such as a series on Technology Management and Environmental Imperatives;
- an annual survey of members' attitudes and spending intentions regarding R&D;
- publication of occasional research reports, focusing on Canadian R&D tax incentives, excellence in the management of innovation, competitive intelligence gathering, and human resources for technology managers.

The International Innovation Institute is a recent addition to the group of private actors in MOT in Canada, based in the Banff Centre for Management, a private not-for-profit education and conference organization in Banff, Alberta. The institute is a "global network of individuals (public and private) committed to understanding processes of innovation and applying this knowledge to the reshaping of their organizations." Its mission is to "assist organizations in restructuring themselves and in reshaping the attitudes, knowledge, and skills of their members in order to nurture high performance in a global, knowledge-based economy." Through networking, conferencing, and educational activities the institute "offers senior managers a new 'window on the world.'" It conducts action research and organizes courses and programs on innovation-driven economies, workplace learning, self-managing teams, managing human resources, effecting change, competitive strategies for advanced technology firms, and preparing organizations for strategic alliances. It also is one of the few applied management education initiatives in Canada to tackle issues of managing innovation at the interorganizational or societal level. Initiatives include direction of a multistakeholder group to produce a strategy for economic development in Alberta.

Mechanisms for Interaction

As indicated previously, the goal of Canadian S&T policy is to reinforce the infrastructure that supports technical change in order to accelerate the diffusion of improved technologies throughout the economy. This raises the issue of the degree to which the improvement of a firm's technological capability requires complementary improvement in the firm's management capability, and how the acquisition of that management capability can be ensured. Answers to these questions depend on how MOT is defined and on how organizational learning processes can be facilitated.

At one end of the spectrum of MOT activity are the many sharply focused technology management techniques and training modules that are readily available on such subjects as leadership in technical teams, project management, R&D management, procurement, and so on. This can be called "technical MOT." Training in technical MOT is targeted at line managers or middle managers. However, as the next section illustrates, Canadian opinion attributes failure to attain competitiveness to a combination of technical deficits and management deficits at the senior level, with the key bottlenecks not in the technological stream but in the organizational-managerial stream (de Vos 1990, 20). The challenge of technological change is increasingly regarded as one of integrative management: "the integration within a company . . . of increasing numbers of functions and resources, both technical and organizational" (ibid. 11). The integrative approach to MOT "is a way of thinking," not merely a collection of techniques, and it is "not easily understood, let alone taught." The integrative task is "not only the main challenge [of managing technical change] but also the most difficult to achieve, and it is the area where most companies fail and where good MOT teachers or consultants are the scarcest" (ibid. 11). The more holistic and integrative the management task, the more it must involve senior management in interaction with all elements in the organization as well as with outside stakeholders.

In Canada most MOT services to firms, publicly or privately delivered, are of the modular, narrowly technical variety. The only current integrative management practices are business planning (i.e., strategic planning) and total quality management (TQM).[8] De Vos (1990, 13) advocates that integrative MOT services be grafted onto business planning or total quality management activities. Some countries are approaching the delivery of MOT services to business via the integrated planning route. Australia, through its National Industrial Extension Service's World Competitive Manufacturing Program, requires firms desiring technical assistance to undergo a three-day integrative business-planning exercise. At the other end of the spectrum, West German agencies focus first on the product or technical problem, eventually embracing broader management practices in some cases. In one U.S. state, government-stimulated improvement of MOT change in firms is regarded as a two-tiered task. One tier concerns the vast majority of firms, which seek technical assistance and technical MOT services for modernization and incremental improvement of performance. The second tier concerns firms with high growth potential in a strategic area ("foundation firms"); these firms are invited to anticipate and confront the organizational and technological challenges to their growth and competitiveness. Sometimes technical assistance relating to the automated production needs of the firm is used to draw some firms into an integrative reformulation of competitive strategy (ibid. 16).

There are special organizational requirements for successfully delivering technical MOT and integrative MOT services to firms from public technical support agencies. Technical MOT is relatively inexpensive, easily disseminated in modular form through diverse kinds of media, such as distance education or videos; it is targeted at line and middle managers and appropriate for incremental modernization of large numbers of firms. Integrative MOT is a type of firm-specific support. It implies substantial interaction between delivery agents and firms. It requires a great deal of commitment and involvement of senior management. Because it is highly customized, it is expensive to deliver. Furthermore, since it can only work in firms that are already dynamic, it amounts to investing in "winners" (de Vos 1990). Finally, integrative MOT requires a special skill set on the part of delivery agents. Public officials, academics, and most business consultants do not have these skills. This implies the necessity of recruiting and training a corps of dedicated delivery agents.[9]

It is likely that quite a few public and academic actors are offering MOT-related technical and management services to firms. No reliable inventory of this activity exists. One example among many is the Ottawa-Carleton Research Institute (OCRI), a creation of universities, community colleges, businesses, and local economic development agencies in the Ottawa region. OCRI represents an innovation-oriented confederation of actors that jointly promotes scientific, technical, and management courses at the metropolitan level.

Summary Profile

Universities, governments, and the private sector have taken a variety of initiatives to develop and disseminate MOT skills and information to constituencies in Canada. The first wave of initiatives took place mainly in universities. A second wave of difficult-to-document initiatives concerns the development of a private market for MOT services. In a third wave that has yet to gather much momentum, public agencies are considering ways to diffuse MOT skills widely throughout the business sector.

The initiative to define the MOT agenda in Canada is currently in the hands of at least five separate groups:

1. the engineering and management schools, which compete with each other for resources to sustain teaching and research programs in MOT and other fields;
2. the international MOT scholarly and practitioner community, which is based in professional associations and in management and administrative disciplines;
3. the federal, provincial, and local governments, which occasionally fund programs in MOT and which are considering ways to deliver technology management services to firms;
4. firms that use MOT skills or services and occasionally support MOT initiatives in-house or in universities;
5. interest groups and policy advisory bodies that compete to articulate a competitiveness agenda, which increasingly identifies management deficits as a problem requiring remedy.

Prevailing Influences on MOT

The idea that managerial incompetence is at the root of uncompetitive firms has become firmly established in the policy debates in Canada since the late 1980s. A 1992 NABST report on technology diffusion in Canada observed that

> the critical factor governing whether a firm remains competitive is the awareness and commitment of the senior managers of that firm. Too

> often, these managers are not aware of the pace of change in competitive firms around the world. They do not take advantage of the support infrastructure and programs which are available to them. They react too late, if at all, to the opportunities of better technology and to the need to change management concepts and procedures to permit their firms to react quickly and responsively to market demands and challenges. This lack of awareness and motivation has been identified as the most serious impediment for [*sic*] technology acquisition and diffusion in Canada. (NABST 1992)

The report of the Prosperity Initiative's consultations on financing innovation provides an even harsher assessment of Canadian management. It observes that "in all quarters, Canadian management was indicted for its overall lack of vision and poor leadership." The indictment blames public, private, labor, and academic management for behaving according to a "herd mentality" and thereby becoming "blockers and followers," and "risk averse, conservative, cautious, ill informed, and insular" (Prosperity 1992d, 17, 29).

These are strong words, delivered by messengers that governments cannot easily silence. They reflect what is probably widespread frustration with Canada's performance, institutions, and leaders. The deluge of reports, briefs, and consultations on Canada's competitiveness has produced a wave of diagnoses of problems and recommended remedies. In the Appendix to this chapter, we list recent ideas and proposals regarding the improvement of innovation management skills. The proposals concern training methods and arrangements, partnerships, new forms of policymaking, recognition and incentives, and new principles.

These proposals represent an agenda for profound change in Canadian management styles, activities, and institutions. They also represent quite divergent diagnoses of problems, thereby generating an agenda that is incoherent. For example, how do we reconcile the call for more cooperation but less herd mentality, better interinstitutional linkages and greater institutional accountability, or more highly specialized training and holistic, visionary leadership?

Contradictory management desiderata are a reflection of our turbulent times. If there is an emergent paradigm for innovation management, it may be that of the "learning organization" that adapts, survives, and grows through nimbleness and intelligence. Characteristics of learning organizations are that they reward critical thinking, initiative, and risk taking; they have clear and committed leadership;

will be a technological one. Northern Telecom, for example, has led the way in producing electronic circuits without the use of chlorofluorocarbons. Not only is this a leading stance environmentally, but it is resulting in dramatic savings (BCSD 1992; Schmidheiny 1992). New materials are allowing goods to be delivered with less, more easily recyclable packaging. Even transportation and distribution systems are being modified to handle goods that are less "packaged" than before. Technology management is fundamentally a skill required to manage change. It makes sense, therefore, that as corporations move into an era of dramatic and fundamental changes, they may turn to the experts on managing change: MOT researchers. We must be ready to meet that challenge.

Corporations in the twenty-first century will find it essential to be able to innovate and operate competitively while at the same time operating in an environmentally responsible manner. The Canadian government and some Canadian corporations have recently taken several strong initiatives in the area of environmental stewardship and sustainable development. MOT research that ties innovation and MOT change to sustainable global competitiveness will go a long way toward ensuring that Canada and Canadians remain healthy and prosperous into the next century.

In summary, there is no simple answer to the question of what are the sources of influence on MOT in Canada. MOT has fuzzy boundaries, and definitions of the content and contours of the field vary. However, the center of gravity is clearly in the engineering and administrative disciplines. As we noted earlier, academic MOT activities are the most highly visible and have received the largest portion of public financial support. Academic traditions and the approaches of the various management and engineering disciplines that make up the core of academic MOT in Canada largely determine the vocabulary and outlook of the "business paradigm" of MOT. In Canada, through the creation of MOT university chairs and new programs devoted especially to the development of business-paradigm MOT research and teaching, funding agencies and universities are primarily mobilizing engineering and administrative disciplines in pursuit of practical solutions to problems encountered within firms. Other social science disciplines offer competing paradigms, but these rarely interact with mainstream business-paradigm MOT. Furthermore, the Canadian style of business-paradigm MOT practice and scholarship is not easily distinguishable from "international" MOT, which is primarily based in North America and, to a lesser degree, Europe.

Most commonly, MOT teaching and research programs have been established in management or engineering faculties. Stand-alone MOT institutions in the university environment are rare. Canadian MOT shares the firm-centered bias that is characteristic of North American management disciplines. MOT appears most accessible and relevant to certain medium and large firms, and to small technology-intensive firms.

A great deal of MOT activity that takes place is difficult to compile, either because it is integrated into other activities, such as promotion of advanced manufacturing technologies, or because it is distributed as sharply focused improved management techniques (total quality management, design, or marketing), often through organizations that do not label themselves as active in MOT.

Although its center of gravity is in the administrative and engineering disciplines, business-paradigm MOT is largely problem- or issue-oriented, without restricting or explicitly favoring particular disciplinary approaches (see the examples in Adler, 1989). Included in the purview of the popular international MOT conference held biannually at the University of Miami are papers on S&T policy, processes of technological innovation, strategic planning, technology transfer, R&D management, entrepreneurship and new venture management, product and process innovations, technological innovation in organizations, quality and productivity issues, MOT in manufacturing and the service industries, managing information technology, human and social issues, education and training issues, economic analysis of innovation, and MOT methodologies. The conference draws participants from a variety of disciplines. However, the conference deals with issues more closely related to engineering management than to social science–based administrative disciplines.

It is important to emphasize that even business-paradigm MOT contains a wide variety of activities, ranging from highly integrated strategy-oriented teaching and research to quite narrowly focused technical training with a management dimension. As de Vos has observed,

> somewhere between these extremes are blends of MOT material with the more established teaching disciplines, such as engineering (including automated design and manufacturing), information technology, or-

> ganization theory, research management, innovation, marketing, total quality management, business policy, and so forth. (de Vos 1990, 39)

Selective investment in the MOT business paradigm in universities has several emerging consequences. One is that problems concerning innovation in environments larger than the firm receive comparatively little attention. Geographers, sociologists, regional economists, and political scientists, rather than researchers in engineering management or business administration, are developing the Canadian knowledge base about innovative environments. Another consequence is that policy issues and problems are underexamined, leaving actors in the policy system to focus on process to the exclusion of content. A third consequence is that uses of technology for purposes that are not immediately economic (e.g., problems of health or environmental technologies) are left for others to examine. Fourth, issues having to do with labor and management-labor relations are not sufficiently brought into the MOT teaching and research stream. Fifth, researchers working on MOT problems outside the business paradigm are scattered throughout departments, faculties, universities, and regions, with few shared institutional structures or dedicated resources to support their work.

Business-paradigm MOT teaching and research makes an explicit commitment to provide useful and reliable skills to firms, building upon and complementing the engineering or management backgrounds of students. The utilitarian expectations of MOT are high, although the field is a hodgepodge of techniques, approaches, beliefs, research programs, and rules of thumb. This is partly because of the novelty of the field and partly because MOT is advancing at three different levels of analysis and aggregation: At one level MOT is a collection of techniques designed to solve relatively narrow sets of problems, such as those of purchasing, product development, or R&D management. At a second level MOT attempts to integrate all relevant considerations into a strategic posture, converging with strategic management. At a third level MOT is concerned with organizational and interorganizational learning.

Policy Implications

We began this chapter with a discussion of technology management, and we end it with the observation that the entire system of public and private management education and practice in Canada is currently being called into question as the country comes to grips with the challenges of competitiveness. Technological innovation, globalization, and economic turbulence are the factors that today's managers must learn to use to competitive advantage. Canadian managers must also grapple with a wide variety of indigenous cultural, social, and institutional factors that may be just noise to traditional management science but that constitute and condition today's decision-making environment.

We have observed that the public, private, and academic sectors are each struggling with the challenges posed by MOT and innovation. The public sector requires competent program managers and is responsible for ensuring the delivery of useful technology management services to firms, but the primary success of public policy for MOT in Canada has been to create support mechanisms for academic research. The private sector is calling for useful MOT skills and services, but neither its advisory bodies nor its purchasing behavior permit clear identification of the management and technical skills that are appropriate at different levels and in different circumstances. The university sector is calling for increased resources and support for MOT teaching and research, raising expectations about performance and utility that must be met.

It is not possible at present to identify best practices for training technology managers in Canada. Only anecdotal evidence is available concerning training methods and outcomes. Little is known about how Canadian firms use persons with formal training in MOT. Even less is known about the labor market for individuals with various combinations of technical and management skills.

The dominant features of the Canadian innovation management and training system are experimentation and diversity. These features are simultaneously a virtue and a liability because of the contradictory expectations that Canada has of its management training activities. In an age of turbulence and in a country as diverse as Canada, it is understandable that no agency or group has a clear overview of activities and no incontrovertible definition of best practice with respect to management of innovation. There is no sharply focused Canadian MOT agenda, nor is there likely to be. On the other hand, dispersion of efforts does not enhance the kind of rapid learning, accountability, and demonstrably effective performance that are widely held objectives.

As we pointed out earlier, conventional wisdom attributes the shortcomings of Canadian innovation not to lack of narrow technical or administrative skills but to the failure of senior management to turn innovation into competitive performance. If there is an emerging consensus on managing technology in Canada, it is probably that "much of the adjustment which has to occur in Canadian firms is based less on technology and more on the way businesses manage themselves" (NABST 1992, 23). Firm-level capacity to maintain a dynamic search for best-practice technology requires that management and labor be skilled and motivated, and that an appropriate, coordinated public innovation support infrastructure be provided.

Sustaining international competitiveness in an open, democratic, high-wage country makes an unusual set of demands on social values and the predominant business culture. In Canadian society, entrepreneurial values are widely shared: a large majority of Canadians believe that hard work pays off, that ability judiciously applied leads to gain, that wealth creation is not a negative-sum game. At the same time, Canadians are reluctant to attribute poverty to lack of individual initiative or to social structure (Fletcher 1992; ISTC 1992). If Canada develops what has been called a "high-growth, low-employment" economic trajectory, successful social adjustment to technological change will require a new commitment to develop social and political skills that are not being examined within the predominant MOT paradigm.

It would be foolhardy to predict the evolution of MOT in Canada over the next decade. More useful would be to specify desirable features of MOT teaching and practice in the country. We consider that a greater degree of interaction and cooperation among Canadian university researchers and teachers of MOT, including sharing of cases and curricula, would be useful. It would also be useful for MOT teachers and private sector associations to devote greater efforts to describing the kinds of management and technical skills required in industry, and to prepare teaching and research material with greater accuracy. We would like to see Canadian industry increase the strength and quality of privately supplied MOT training and consulting services through the development of a more vigorous market for these services. We would also like to see a more vigorous and comprehensive set of technical and management services provided to firms by public institutions. Finally, we would like the MOT field to become more reflexive, to question itself, its assumptions, its parameters, and its performance with a greater degree of clarity than has been the case. We would not like to see any one of the three streams of MOT that we have identified—technical MOT, integrative MOT, and learning-organization MOT—suppress either of the other streams, but we would like to see greater involvement of the social science and natural science disciplines in MOT teaching and practice. Finally, we suspect that management thought and practice will be increasingly concerned with environmental issues. These are likely to engage all three streams of MOT, from pragmatic quality and efficiency improvement that reduce harmful environmental impact to the creation and maintenance of the socially responsible "green corporation," in touch with its social, physical, and biological environment.

Appendix: Recent Proposals to Strengthen Management of Innovation in Canada

1. ***Strengthen programs to support the hiring of technically trained people by firms***
   - An estimated 70 percent of Canadian manufacturing firms have no engineers on staff. Some provinces, notably Quebec and Ontario, have programs that subsidize the hiring of technically trained people. Recommendations have been made to expand these programs (SCC 1988; CSTQ 1990).
2. ***Increase the influence of persons with technical competence within senior management***
   - Canadian firms are urged to modify the composition of boards of directors and strengthen the technical capability of senior management to provide better strategic direction with respect to technology (SCC 1988).
3. ***Establish alliances and partnerships***
   - Large indigenous technology firms should provide mentorship to smaller companies in the innovation, production, management, and international marketing of products (CIAR 1988).
   - Canadian firms are urged to sponsor and encourage "all forms of partnerships and alliances" (Prosperity 1992d, 30).
   - Organizations are encouraged to train executives as a team (Larson 1992, 21).
4. ***Establish new institutional arrangements to provide analysis, generate consensus, guide collective action, fund research, and keep innovation and competitiveness issues in the public eye***
   - Recent proposals include: a Prosperity Council to include representatives of federal and provincial governments, labor, education, con-

sumers, and voluntary and social groups (Prosperity 1992a, 62); a Competitiveness Council (Prosperity 1992b); a Centre of Excellence for Sustainable Development (Prosperity 1992a); a Canadian Education Council (ibid.); a Technology Change Centre (ibid.); a Continuous Learning Board (NABST 1991b); a Management Research Council (CAC 1992, 46); a "series of community-based business support networks or planning councils" (Prosperity 1992d, 30); a National Quality Institute (Prosperity 1992a); a national business-leadership-skills school (Prosperity 1992a, 30).

5. ***Improve management training***

- Courses in management of technological change and new product development should be developed and delivered through secondary schools, technical institutes, and universities (Prosperity 1992a, 20). Firms should collaborate with universities to strengthen their teaching and research capacity in MOT (SCC 1988). Business and engineering faculties should work together to develop graduate and undergraduate courses in MOT (CAC 1992, 41).
- Public funding of university-based management programs should be increased and tied to accountability for improved teaching and research (ibid. 44–45, 53).
- Business faculties and industry should work more closely together to improve the relevance and quality of business education (ibid. 40).
- Federal and provincial governments should fund a program to enable graduate engineers to learn designated foreign languages and study abroad (SCC 1988, 12).
- Companies should maintain internal technology management programs (ibid). The goal should be to develop "managers with the confidence, knowledge and skills to manage continuous innovation" (NABST 1991b, 17). Industrial managers should upgrade their skills and those of their employees "through on-going training programs and active collaboration with the education sector" (NABST 1992).
- Universities should improve their continuing education in MOT (SCC 1988). Canadian public and private organizations should increase commitment to continuing education of managers (CAC 1992, 44; Larson 1992).
- Public and private organizations should experiment with new ways of educating managers (CAC 1992, 45).

- Provincial governments should establish programs to allow universities to hire domestic or foreign MOT practitioners for short periods (SCC 1988).
- A statement of competence-based qualifications should be developed for Canadian managers and used to redesign postsecondary management and administrative programs and in hiring and promoting employees (Prosperity 1992a, 46).
- Courses in the management and use of computers and other information and communications technologies should be prepared for all students in management and administration (ibid, 31).
- Programs of mentoring and interchange of managers, executives, labor representatives, and academics should be established to accelerate the diffusion of new management practices (Prosperity 1992a, 46; NABST 1991b, 17).
- A total-quality-management education campaign for CEOs and senior management should be launched.

6. ***Improve labor-management teamwork***
   - Labor-management teamwork should be adopted as "the normal way of doing business in Canada" (Prosperity 1992a, 31).
   - Models of labor-management teamwork should be developed and promoted (ibid.), and training for such labor-management collaboration should be provided for total quality management (Hutton 1992).
   - Labor-management cooperation should be promoted "by building national forums for industry associations to pursue common objectives" (Prosperity 1992a, 31).
7. ***Develop management innovation strategy***
   - Corporate innovation should be fostered through total quality management programs (FCSA 1992, 17).
   - Senior management in firms, trade associations, financial institutions, and labor organizations should "make a point of acquiring enough operational knowledge about the strategic management of technological change to provide effective management and vision" (SCC 1989, 17). Trade associations should help "de-mystify the role of technology in business operations, . . . nurture greater commitment on the part of CEOs toward employing new technology and managing technology more effectively," and "develop a better understanding of technological innovation in business planning" (CCC 1988).

- Firms "are on the front lines of international competition." The burden is on their managers to understand their competitive position and focus on innovation-driven advantage (Porter 1991, 77).

8. ***Offer recognition and tax incentives***
   - Create "one-time, tax-free rewards for achievement in business excellence" (Prosperity 1992d, 30).
   - Government should provide tax incentives for people to serve on SME boards of directors or advisory committees (ibid.).
   - Firms should recruit offshore Canadians and expatriate foreigners with tax-free bonuses and incentives (ibid.).
   - Government should establish new tax incentives and other inducements to promote technology acquisition.
9. ***Adopt improved management practices***
   - Industrial managers should practice continuous benchmarking, and government programs and private technology centers should actively promote the identification and diffusion of best practice in this respect (NABST 1992).

Notes

1. The value of the Canadian dollar in recent years was approximately U.S.$0.75–0.80.

2. Forty-five recent reports and studies are described in Prosperity (1992c).

3. In so doing, it shed light on one of the most peculiar dimensions of the Canadian national system of innovation: significant parts of this system are located in the United States.

4. For a description of several of these incentives and programs see Botham and Giguère (1993).

5. The preface to the Prosperity Initiative action plan urges citizens to read the plan and discuss it with friends and colleagues and adds, "Wherever you can support this plan with personal action, we urge you to take that action" (Prosperity 1992a, i).

6. On the limits of this model of development see Phillips (1991).

7. The seminars were: "Made in Canada: Using Manufacturing to Gain a Competitive Edge," "Winning at New Products," "Effective Technology Transfer," "Bridging the Authority Gap: Influence Strategies for Project Managers," "The Management of Technology" (a special seminar for CEOs and executives), "Building Blocks for Teams: How to Lead Effective Groups," "Development of First-Line Supervisors," "Effective Employee Suggestion Systems," "Profit from Information Technology," "Managing Development Projects," "Managing Strategic Change in Continuous Operations Industries," and "Meeting the Technology Challenge: From Initiative to Implementations."

8. A statement of TQM's principles shows why it is integrative. TQM is led by top management; it is customer-focused and aims at continuous improvement

to attain long-term company goals. It involves everyone in a search process for win-win solutions. It is prevention-oriented, systematic, and methodical; it is based on management by fact, and it promotes the view that companies are responsible to society (Hutton 1992).

9. According to de Vos (1990, 38), the ideal teacher-consultant for delivering integrative MOT services has a technical degree, graduate business education, several years' experience in technical management and other business experience, and good communication and facilitation skills. The consultant may be a generalist with depth in one or two core disciplines, has rational and intuitive problem-solving capabilities, inspires cooperation, can listen and synthesize well, has linkages with an investment firm, has experienced and overcome failure, and is either around thirty-five years old or is an older person with about twenty-five years of experience.

10. In an advertisement accompanying the article, P&G states: "Total Quality Management is not just blue sky. It can also be cleaner air, clearer water and greener earth. Making our world a better home for people today, and for future generations, has become an important concern of consumers, customers and communities all over the world. Here and overseas, industry is starting to move away from merely meeting governmental regulations and toward a process of continuous environmental improvement. That's why the application of Total Quality Management to the environment has created so much interest in the business community. At Proctor & Gamble, we are learning how to use TQM to develop practical new solutions to environmental issues related to our processes, products and packaging. We're eager to share our experiences with others who want to contribute to progress in achieving environmental goals."

References

Abbott, Mark. 1988. "The Need for Education in Management of Technology in Canada." Paper presented at the Administrative Sciences Association of Canada (ASAC) Conference, Halifax, Nova Scotia.

Adler, Paul S. 1989. "Technology Strategy: A Guide to the Literatures." In *Research on Technological Innovation, Management and Policy*, ed. R. S. Rosenbloom and R. A. Burgelman, 25–151. Vol. 4. Greenwich, CT, and London: JAI Press.

Anderson, Frances, and Charles H. Davis. 1993. "The Government Research Infrastructure." In *Science and Technology in Canada*, ed. P. Dufour and J. de la Mothe, 47–660. London: Longman.

APRO (Association of Provincial Research Organization). 1992. *Advanced Manufacturing Technologies in Canada*. Unpublished report. Ottawa.

ASQC (American Society for Quality Control). 1992. "Quality '92: Leading the World-Class Company." *Fortune* 126(6): 158–223.

BCSD (Business Council for Sustainable Development). 1992. *Report on Technology Cooperation: Technology Advisory Group Working Paper*.

Blais, André. 1986. "The Debate on Canadian Industrial Policy." In *Industrial Policy*, ed. A. Blais, 55–820. Toronto: University of Toronto Press.

Botham, Richard, and Michel Giguère. 1993. "The Industrial Research Infra-

Committee on Science Policy. Vol. 3, *A Government Organization for the Seventies.* Ottawa: Information Canada.

Larson, Peter E. 1992. *Management and Executive Development in a Changing World: A Research Report Prepared for the Prosperity Secretariat.* Ottawa: Prosperity Initiative, Industry, Science and Technology Canada.

Lefebvre, Louis, Elisabeth Lefebvre, and Anne Le Luel. 1992. *La gestion de la technologie: un choix ou une nécessité?* Montreal, Quebec: Le Centre de développement technologique, École Polytechnique.

Leonard-Barton, Dorothy. 1992. "Management of Technology and Moose on Tables." *Organization Science* 3(4); 556–58.

Leslie, Peter. 1987. *Federal State, National Economy.* Toronto: University of Toronto Press.

MTI (Management of Technology and Innovation Institute). 1988. *Proceedings of the Second National Workshop on Management and Technology: The Education and Development of Tomorrow's Industrial Leaders.* Hamilton, Ontario: Management of Technology and Innovation Institute.

MTI. 1990. *1990 Fall Calendar Management Development Programs.* Hamilton, Ontario: Management of Technology and Innovation Institute.

NABST (National Advisory Board on Science and Technology). 1988. *Industry Committee Report. Presented to the Prime Minister of Canada.* Ottawa: National Advisory Board on Science and Technology.

NABST. 1991a. *Science and Technology, Innovation and National Prosperity: The Need for Canada to Change Course.* (April). Ottawa: National Advisory Board on Science and Technology.

NABST. 1991b. *Learning to Win: Education, Training and National Prosperity.* Report of the Human Resource Development Committee, April. Ottawa: National Advisory Board on Science and Technology.

NABST. 1992. *Measuring Up to the Benchmark and Moving Ahead.* Report of the NABST Committee on Technology Acquisition and Diffusion. Ottawa: National Advisory Board on Science and Technology.

Niosi, Jorge. 1991. "Canada's National System of Innovation." *Science and Public Policy* 18(2): 83–92.

NRC (National Research Council). 1987. *Management of Technology: The Hidden Competitive Advantage.* Washington, DC: National Academy Press.

OECD (Organization for Economic Cooperation and Development). 1990. *Advanced Materials: Policies and Technological Challenges.* Paris: Organization for Economic Cooperation and Development.

OECD. 1992. *OECD Economic Surveys: Canada.* Paris: Organization for Economic Cooperation and Development.

Ottman, J. A. 1992. "Industry's Response to Green Consumerism." *Journal of Business Strategy* 13(4): 3–7.

Phillips, Paul. 1991. "New Staples and Mega-Projects: Reaching the Limits to Sustainable Development." In *The New Era of Global Competition: State Policy and Market Power*, ed. Daniel Drache and Meric S. Gertler, 229–46. Montreal and Kingston, Ontario: McGill-Queen's University Press.

Porter, Michael E. 1991. *Canada at the Crossroads: The Reality of a New Competitive Environment.* Ottawa: Business Council on National Issues and Ministry of Supply and Services.

Prosperity. 1992a. *Inventing Our Future: An Action Plan for Canada's Prosperity*. Ottawa: Steering Group on Prosperity.

Prosperity. 1992b. *Prosperity through Innovation: The Task Force on Challenges in Science, Technology and Related Skills*. Background report. Ottawa: Prosperity Initiative; Industry, Science and Technology Canada.

Prosperity. 1992c. *An Overview of Selected Studies on Canada's Prosperity and Competitiveness*. Ottawa: Prosperity Initiative; Industry, Science and Technology Canada.

Prosperity. 1992d. *Financing Investment: The Consultation Process—Overview and Final Recommendations*. Ottawa: Prosperity Initiative; Industry, Science and Technology Canada.

Roobeek, Annemieke J. M. 1990. *Beyond the Technology Race: An Analysis of Technology Policy in Seven Industrial Countries*. Amsterdam: Elsevier Publishers.

Salter, Liora. 1990. "Overview of the Field [of Management of Technical Change] and its Researchers [in Canada]." In *Managing Technology: Social Science Perspectives*, ed. Liora Salter and David Wolfe, 17–43. Toronto: Garamond Press.

Salter, Liora, and David Wolfe, eds. 1990. *Managing Technology: Social Science Perspectives*. Toronto: Garamond Press.

SCC (Science Council of Canada). 1988. *Gearing Up for Global Markets: From Industry Challenge to Industry Commitment*. Ottawa: Science Council of Canada.

SCC. 1992. *Reaching for Tomorrow: Science and Technology Policy in Canada, 1991*. Ottawa: Science Council of Canada.

SCC. 1989. *Enabling Technologies: Springboard for a Competitive Future*. Ottawa: Science Council of Canada.

Schmidheiny, Stephan. 1992. *Changing Course: A Global Perspective on Development and the Environment*. Cambridge, MA: MIT Press.

Weimer, William. 1991. "Education for Technology Management." *Research/Technology Management* (May–June): 40–45.

West, Alan. 1992. *Innovation Strategy*. New York: Prentice Hall.

3

Developing Technology Managers in Singapore: Issues and Challenges

Wong Poh-Kam

Background

Technology has increasingly been viewed by the newly industrializing economies (NIEs) as the means for catching up with the industrialized nations. While early on this catch-up effort has been concentrated on developing the scientific and engineering workforce to absorb and assimilate technologies imported and transferred from the more advanced industrial countries, there is increasing recognition by government policy makers and private business enterprises of the importance of management of technology (MOT)—the effective management of the process of technological utilization and development to achieve higher productivity and greater competitiveness. The concern with improving the technology management process leads naturally to the question of how to improve the quantity and quality of technology managers. This, in turn, leads to the question of which skills are essential to technology management, and how such skills can be more effectively provided.

The demand for MOT skills can be met in principle through a variety of supply mechanisms or skill-formation mechanisms, including temporary expatriate staffing or permanent in-migration of foreign technology managers, formal training programs either at local or overseas universities and training institutions, corporate in-house training and/or learning-by-doing, and various continuing education programs, such as short executive programs, workshops, and conferences. These different skill-formation mechanisms often address different MOT skill needs, although some of them are close substitutes for one another. There may also be significant complementary relationships between some of these mechanisms. In particular, it has been argued by many that the classroom learning style of formal M.B.A. programs is often inadequate and needs to be complemented with on-the-job learning and direct managerial experience dealing with particular technologies in the context of a specific industry.

There is currently no explicit theory of the relative effectiveness of the different skill-formation mechanisms, and still less consensus on what the optimal composition of mechanisms should be for a country at a particular stage of development with a particular pattern of demand for skills. However, through case studies of the experience of different countries at different levels of economic development, some generalizations may be possible. With this in mind, a review of the experience of one NIE, Singapore, may be instructive.

Like the other Asian dragons, Singapore has achieved remarkable economic growth over the last three decades. As can be seen from Table 3.1, Singapore's real GDP grew at an average of 8.1 percent per annum over 1960–92, while its GNP per capita (in current prices) increased from S$1,330 (U.S.$435) in 1960 to S$27,028 (U.S.$16,582) in 1992.

Technology has played a critical role in this rapid industrialization of the Singaporean economy. An examination of how MOT skill formation has evolved over the years to meet the needs of economic development in Singapore may prove useful to other developing economies. Moreover, in the future, Singapore's economy will become increasingly high-tech-driven. Under the Strategic Economic Plan (SEP) and the National Technology Plan (NTP) formulated in the early 1990s, the Singapore government's strategic intent in the 1990s is to further promote selected high-tech industry clusters and to accelerate research and development (R&D) and other technology-intensive activities. Besides accelerating the demand for a scientific and technological workforce in general, such development will further increase the demand for technology management skills both in the private and public sectors. The MOT issues and challenges that Singapore is currently facing may therefore serve as good indicators of the likely challenges other developing economies may face as well.

Table 3.1

Economic Development Profile of Singapore

	1960	1992
Population (million)	1.5	2.8
Area (square kilometers)	581	641
Average GDP growth per annum, 1960–92, as a percent:	{8.1}	
Per capita GNP (S$)	1,330 (U.S.$435)*	21,271 (U.S.$13,050)*
Gross national savings (as a percentage of GNP)	–2.6	46.4
Total trade (S$ billion)	7.6	220.9
Total exports (S$ billion)	3.5	103.3
Overall balance of payments (S$ billion)	0.14	9.9
Unemployment rate (as percent of labor force)	14	2.7
Percentage of population in public housing (about 80% of this is privately owned)	9	87
Infant mortality rate (per 1,000 live births)	34.9	5.0

Sources: Economic Development of Singapore 1960–1991; Yearbook of Statistics Singapore 1992.

Note: The figures for 1992 are preliminary figures.

*Exchange rate in 1960: U.S.$ = S$3.06; in 1992: U.S.$ = S$1.63.

This chapter has three objectives. First, it provides a broad overview of the changing role of technology in Singapore in the context of rapid industrial development over the last three decades, and the impact these changes have on the demand for management of technology. Second, it highlights the emerging MOT challenges that Singapore faces in the 1990s. Third, it analyzes the significant initiatives undertaken by the government, the universities, and private industries in recent years to improve the supply of technology managers to cope with these challenges.

Overview of Development Strategy and the Changing Role of Technology

In a nutshell, in the last thirty years Singapore's rapid economic growth has transformed it from a labor-surplus, entrepôt-trade-based urban mercantile center within a largely rural, regional hinterland into a high-skill-based, modern regional business hub involving high value-added manufacturing, transportation, trading, financial, and business services within one of the fastest growing hinterland regions (ASEAN) in the world. In achieving this remarkable transformation, government economic development strategy has evolved as circumstances changed. To facilitate discussion, we can divide this transformation process into four phases (Wong 1994).

Phase I: Early Industrialization Phase (1959 to Late 1960s)

In the period from independence in 1959 to the late 1960s, the Singapore government's most important concern was employment creation. This was accomplished through a concerted effort to attract foreign investment in export-oriented manufacturing activities that created jobs. To make Singapore attractive for such investment, the government embarked right from the start on a program of infrastructural developments, including industrial estates, public utilities, telecommunications, and seaport facilities. At the same time, the government correctly recognized that the major bottleneck at that early stage of economic development was not caused by too few skilled workers, but by a lack of experienced managers and plant engineers. Consequently, a liberal policy of free immigration of foreign managers and technical staff was introduced to allow multinational corporations (MNCs) investing in Singapore to bring in the necessary personnel from overseas.

Phase II: Technological Upgrading (Early to Late 1970s)

The foreign-investment-promotion program was highly successful and attracted major investments by MNCs, especially in the electronics sector. At the same time, the favorable location of Singapore also started to attract major investments in ship repair and maintenance and in petroleum refining. With more manufacturing activity in Singapore, its labor surplus turned into a labor shortage, and an increasing demand for local supporting industries in machining, mold-and-die making, and precision engineering developed. To meet the rising demand for skilled workers, technicians, and technically competent supporting industries, the government embarked on a massive program in industrial training to upgrade the skill levels of the workers and to increase the supply of technicians and engineers. Infrastructural investments were also intensified, particularly in transport facilities and telecommunications.

Phase III: Industrial Restructuring and Economic Diversification (Late 1970s to Late 1980s)

By the late 1970s, Singapore's economy had experienced a significant labor shortage. To meet this challenge, the government initially drew upon a ready supply of labor from Malaysia. However, as the number of foreign laborers became too large, the government embarked on a program to restrict the influx of foreign workers and to introduce a deliberate high-wage policy to force firms to upgrade and phase out the lower-skilled, labor-intensive activities.

The industrial restructuring and high-wage policy of the late 1970s continued over the first half of the 1980s. At the same time, financial and business services were emphasized as part of the industrial restructuring to move the economy away from too high a dependence on manufacturing.

Partly as a result of slowdowns in the global and ASEAN regional economies and partly exacerbated by firms' failure to achieve productivity growth in line with the high-wage policy and the rapid increase in land costs, the Singaporean economy lost competitiveness versus other Asian NIEs and experienced a sharp recession in 1985. This prompted a major review of existing government policies and the formulation of a new economic development strategy to further diversify the economy and to improve its resilience (Ministry of Trade and Industry 1986). The high-wage policy was moderated, and a flexi-wage system was put into practice. The industrial restructuring that started in the late 1970s was continued, but with greater emphasis on government programs and policies to help local enterprises, especially the small and medium-size enterprises (SMEs), to grow and upgrade their technological capabilities. Business growth and technology transfer through linkages between local SMEs and the large foreign MNCs were also given greater emphasis. The concept of strategic business partnership was promoted to encourage MNCs to make long-term commitments to operate in Singapore through continuous upgrading. New technology-intensive industries, such as aerospace industries, information technologies, and printing/publishing, were promoted, while the business services, telecommunications, and transportation/logistics

Table 3.2

Singapore's GDP Distribution by Sector, 1960–92
(as a percentage)

Industry	1960	1970	1980	1990	1992
Agriculture and mining	3.9	2.7	1.5	0.4	0.3
Manufacturing	11.7	20.2	28.1	28.0	26.6
Utilities	2.4	2.6	2.1	1.9	1.7
Construction	3.5	6.8	6.2	5.4	7.5
Commerce	33.0	27.4	20.9	16.3	15.1
Transport and communication	13.6	10.7	13.5	12.5	13.1
Financial and business services	14.4	16.7	18.9	25.5	25.7
Other services	17.6	12.9	8.7	9.9	9.9
Total	100%	100%	100%	100%	100%

Sources: Calculated from MTI (1990). Yearbook of Statistics Singapore (1992).
Note: Figures may not add up to 100 percent due to rounding. Figures for 1992 are preliminary figures.

support sectors were given greater emphasis in order to strengthen the competitive advantage of Singapore's infrastructures.

The changes in Singapore's economy over these three phases of growth are clearly reflected in the evolving sectoral structure of Singapore (see Table 3.2). The manufacturing sector expanded rapidly in the 1970s and then stabilized in the 1980s at around 28–30 percent of total GDP. The financial/business services sector expanded its share gradually in the 1960s and 1970s and then grew rapidly in the 1980s. The transport and communication sectors' share of total GDP, after a slight decline in the 1960s, rose back to around 13 percent by 1980 and has maintained that proportion ever since.

Emerging Phase: Total Business Center, Knowledge-Intensive Industrialization, and Internationalization (1990s)

The end of 1990 saw the transition of political leadership from Lee Kuan Yew, who had been at the helm since independence, to a new and younger generation of political leaders headed by Goh Chok Thong. This transition helped crystallize the explicit formulation of a more coherent, comprehensive vision for the long-term strategic development of Singapore by the political leadership. The goal of this new vision, called "the Next Lap," is to enable Singapore to attain the status and characteristics of a first league developed country within the next twenty to thirty years (Government of Singapore 1991). To achieve this vision, a Strategic Economic Plan (SEP) was prepared by the Ministry of Trade and Industry (MTI) to chart the fundamental economic development policies for the 1990s and beyond (Ministry of Trade and Industry 1991). The key elements of the SEP include: positioning Singapore as a global city or "total business hub" in the Asia-Pacific region on a par with other leading global cities in the world; emphasis on attracting high-tech, knowledge-intensive industries and promoting high-value innovative/creative activities; intensified investment in developing infrastructures and enhancing human resources; promoting teamwork and cooperation between labor, business, and government; and internationalization of local firms as a means to transcend the constraints of small size and to achieve better access to overseas markets and technology. As part of this long-term internationalization strategy, a program to jointly promote the development of a "growth triangle" involving Singapore, Southern Peninsular Malaysia, and the Riau Islands of Indonesia was also initiated to facilitate the relocation of Singapore-based firms to other members of the triangle.

Public Sector MOT Organization

The Singapore government's approach to the development of science and technology (S&T) can best be characterized by its flexible and pragmatic adaptation to the requirements of economic development (Wong 1994). Indeed, unlike many developing countries, Singapore until very recently did not have a well-developed and specialized institutional structure for formulating and implementing S&T policies. In 1968, a Ministry of Science and Technology was formed, but it never played an important role in development strategy and was dissolved in 1981, with the bulk of its S&T functions taken over by the Singapore Science Council (SSC), which had existed since 1967. It was not until 1991, when the SSC was revamped to become the National Science and Technology Board (NSTB) under the Ministry of Trade and Industry (MTI), that an integrated, organizational framework was put in place to deal with S&T strategy formulation and implementation. Even so, the NSTB is considerably smaller in workforce and resource allocation than its counterpart in countries like Korea and Taiwan.

Similarly, until 1991, when a National Technology Plan (NTP) was formulated by the newly established NSTB, Singapore had no formal S&T policy explicitly described in a plan or blueprint (Wong 1994). It is true that as part of its industrial restructuring plan initiated in 1979, the Ministry of Finance did formulate an indicative long-term R&D plan in conjunction with other interested agencies like MTI, SSC, and the Singapore Institute of Standards and Industrial Research (SISIR). Similarly, in 1986, when the "New Directions" policy document was prepared by MTI to help guide the Singaporean economy out of the 1985 recession, there was significant focus on specific S&T-related policies. Nonetheless, these indicative plans and policies were always regarded as part of an overall economic development strategy, and a multiagency framework was always adopted to implement the various policies.

Because of the constantly evolving nature of Singapore's technology development policies, it is somewhat misleading to talk in terms of a single set of public S&T policies that is applicable throughout the past economic development phases (Wong 1994). Nonetheless, the following generic components of technology development appear to have been emphasized by the Singapore government:

- reliance on foreign direct investment (FDI) as a supplier of technology;
- flexible and responsive workforce development programs and policies;
- heavy commitment on infrastructure development to support technological deployment;
- rapid deployment of technology in the public sector itself;
- relatively low emphasis on R&D until recent years.

A more detailed account of how each of the above policy thrusts has been implemented in each of the economic development phases outlined earlier can be found in Wong (1994). For the purpose of this chapter, it is more useful to focus on how these policy components influenced the historical pattern of technological skills formation in general and MOT skills formation in particular.

Table 3.3b

Growth in Administrative/Managerial and Professional/Technical Occupations, 1980

	1980				
		Professional and Technical Workers		Administrative and Managerial Workers	
Industry sectors	Total	Number	%	Number	%
Manufacturing	312,641	14,314	15.33	11,698	22.66
Construction	38,141	3,254	3.48	6,776	13.12
Commerce	244,696	3,007	3.22	8,918	17.27
Transport, Storage, and Communications	122,420	10,112	10.83	5,169	10.01
Financing, Insurance, Real Estate, and Business Services	79,067	11,122	11.91	11,142	21.58
Community, Social, and Personnel Services	222,783	49,635	53.15	7,250	14.04
Other	29,184	1,936	2.07	680	1.32
Total	1,068,932	93,380	100	51,633	100

% of professional and managerial workers in each sector	Professional and Technical Workers	Administrative and Managerial Workers
Manufacturing	4.58	3.74
Construction	5.60	11.65
Commerce	1.23	3.64
Transport, Storage, and Communications	8.26	4.22
Financing, Insurance, Real Estate, and Business Services	14.07	14.09
Community, Social, and Personnel Services	22.28	3.25
Other	–6.63	2.33
Total	8.74	4.83

Source: Report on the Labour Force Survey of Singapore, 1980.

Table 3.3c

Growth in Administrative/Managerial and Professional/Technical Occupations, 1992

	1992				
		Professional and Technical Workers		Administrative and Managerial Workers	
Industry	Total	Number	%	Number	%
Manufacturing	434,121	67,498	21.91	33,992	21.57
Construction	103,222	10,875	3.53	15,037	9.54
Commerce	335,990	29,952	9.72	69,898	44.35
Transport, Storage, and Communications	158,387	25,547	8.29	10,784	6.84
Financing, Insurance, Real Estate, and Business Services	171,449	68,835	22.35	19,836	12.59
Community, Social, and Personnel Services	339,192	102,371	33.23	7,412	4.70
Other	13,792	2,976	0.97	638	0.40
Total	1,576,151	308,055	100	157,397	100

% of professional and managerial workers in each sector	Professional and Technical Workers	Administrative and Managerial Workers
Manufacturing	15.55	7.83
Construction	10.54	14.57
Commerce	8.41	19.61
Transport, Storage, and Communications	16.13	6.81
Financing, Insurance, Real Estate, and Business Services	40.15	11.57
Community, Social, and Personnel Services	30.18	2.19
Other	21.58	4.63
Total	19.54	10.00

Source: Report on the Labour Force Survey of Singapore, 1992.

Table 3.4

Engineering Science Graduates from Tertiary Institutes in Singapore, 1980–92

Graduates from:	1980	1981	1982	1983	1984	1985	1986	1987	1988	1989	1990	1991	1992
Engineering													
National University of Singapore	309	380	413	513	733	348	407	398	487	501	432	498	561
Nanyang Technological University	—	—	—	—	—	557	661	664	695	779	761	880	1,044
Science													
National University of Singapore	291	466	612	707	777	910	1,004	1,074	1,065	1,262	1,398	1,479	1,471
Total university graduates in science/engineering	600	846	1,025	1,220	1,510	1,815	2,072	2,136	2,247	2,542	2,591	2,857	3,076
Polytechnics:													
Engineering													
Singapore Polytechnic	1,651	1,729	1,891	1,923	2,076	2,246	2,557	3,059	3,052	3,120	3,434	3,268	3,363
Ngee Ann Polytechnic	360	490	629	544	707	824	999	1,829	1,987	1,815	1,881	2,022	2,230
Computer studies													
Singapore Polytechnic	—	—	—	—	—	—	—	134	130	154	158	191	282
Ngee Ann Polytechnic	—	—	—	—	105	103	193	199	193	199	187	249	254
Total polytechnic graduates in science/engineering	2,011	2,219	2,520	2,467	2,888	3,173	3,749	5,221	5,362	5,288	5,660	5,730	6,129
Institute of Systems Science													
Diploma in knowledge of English	—	—	—	—	—	—	—	—	—	—	24	25	16
Diploma in systems analysis	—	—	32	42	43	45	49	49	52	50	49	53	49
Total tertiary institute graduates in Singapore	2,611	3,065	3,577	3,729	4,441	5,033	5,870	7,406	7,661	7,880	8,324	8,665	9,270

Source: Singapore Yearbook of Labour Statistics, various years.

in the proportion in S&T personnel increasing significantly from 31 percent in 1980 to 42 percent in 1992.

In addition to the increasing domestic supply of S&T graduates, a significant proportion of Singapore's stock of university graduates in S&T is derived from overseas (either from Singaporean students studying S&T fields overseas or immigrants with S&T qualifications). In 1990 the number of Singaporean students studying in universities overseas was slightly more than half the number of Singaporean students in local universities. Moreover, there tends to be a larger proportion of overseas university graduates in S&T-related fields; in 1990, 44.5 percent of employees who were university graduates in S&T studied overseas compared with 38.2 percent of all employees with university degrees (Singapore Department of Statistics 1992).

The number of degree holders in science and technology does not tell the whole story. In addition there has been a significant increase over the years in the number of polytechnic graduates with diplomas in various S&T fields, although the growth rate has been lower than that for university graduates over the last decade (Table 3.4). Indeed, it is the ability of the Singapore government to generate a large number of competent technicians at the polytechnic level in the early phases of economic development that distinguishes Singapore from most other developing countries; it is only in the more recent years, when the Singaporean economy became more advanced, that the production of university graduates became a higher priority.

R&D Workforce

Another indication of the changing pattern of technology skills in the Singaporean economy is the rapid rise in the number of people involved in R&D in general and the number of research scientists and engineers (RSEs) in particular. In absolute terms, the number of RSEs increased from about 1,200 in 1982 to 6,500 in 1992. As a proportion, RSEs per 10,000 workers in the labor force increased from 10.6 to 39.8 over the same period.

Pattern of MOT Skills Formation

Disaggregated statistics are not available to indicate how much of the managerial and S&T workforce employed in the Singaporean economy falls within the broad definition of technology management. A very rough guesstimate is that up to one-fifth of the managerial positions could be regarded as having a high MOT content, and perhaps up to one-eighth of those in the professional/technical categories. What is clear, however, is that the nature and composition of MOT skills required of technology managers has changed drastically over the years. The following stylized account serves to highlight the major changes over the different economic development phases outlined earlier.

Early Industrialization Phase

In the first phase of industrialization, the critical personnel with the MOT skills required to get the industrialization process started were experienced production managers for the new manufacturing plants started by MNCs. Given the shortage of such managers in Singapore, the government wisely adopted a liberal immigration policy to allow the MNCs to supply these experienced managers, who effectively transferred their experience and management know-how to Singapore. Indeed, as observed by Dr. Goh Keng Swee, the main architect for the early industrialization strategy of Singapore, these expatriate managers also played an effective entrepreneurial role by taking the considerable risk involved in the attempt to create a system of industrial management in a country where none existed before. The Singapore government's main contribution to MOT at this stage, besides the liberal expatriate staffing policy, was the training of industrial supervisors and technicians to staff the lower level of operations management.

Technological Upgrading Phase

While the management of the transfer of basic production processes from overseas to Singapore was the focus of the initial industrialization phase, the emphasis shifted toward expanding middle management know-how in upgrading and improving the production process technology in the subsequent phases of technological development. Although the presence of experienced expatriate production managers continued to be important, there was an increasing number of local industrial engineers being trained, largely on the job, to take on production management responsibilities. At the same time, the growing importance of local subcontracting by the MNCs resulted in a fairly large increase in the number of small and medium-size enterprises (SMEs) providing a wide range of services such as machining, precision engineering, mold and die making, and metal stamping, which support local industries. Many of these SMEs were started by technicians and skilled workers previously working in MNCs. However, these entrepreneurs typically lacked formal training in management skills.

In this phase the government played an indirect MOT role, sending the brightest students and high-flying civil servants to study engineering, science, or management/public administration at top universities overseas and promoting rapid expansion in the training of engineers and technicians at the university and polytechnic level. Through their employment in manufacturing and other technology-based enterprises, such as construction, shipbuilding/repair, and public utilities, many of these engineers and technicians were promoted into lower- and middle-management positions. Some of these positions were also filled by graduates from the School of Management at the National University of Singapore (NUS), while private institutions, such as the Singapore Institute of Management (SIM), also provided management training through various diploma and certificate programs.

The role of the private sector in MOT skill formation was largely to provide on-the-job training, although some of the large MNCs also provided significant formal, structured, in-house corporate training.

Economic Diversification Phase

The technological upgrading process continued into the third phase of the economic development of Singapore, with increasing emphasis being placed on factory automation and the introduction of advanced manufacturing technologies (AMTs) such as robotics, computer-numeric control (CNC), and CAD/CAM, since the mid 1980s. However, an even greater emphasis was made during this phase to diversify the technological deployment process to other economic sectors, especially finance, trade, logistics, and business services. In particular, the diffusion of information technology became much more important during this period. Under the initiative of the National Information Technology Plan, Singapore has achieved remarkable growth in information technology (IT) usage in both the private and public sectors in the 1980s. The rapid diffusion in the public sector has been achieved through a concerted Civil Service Computerization Program (CSCP) launched in the mid 1980s. Between 1985 and 1990, the number of mainframe computers and minicomputers installed in government departments jumped from 35 to 107, the number of systems developed increased from 72 to 293, and annual spending on computers rose from S$14 million to S$200 million. Major statutory bodies in Singapore, such as the Port Authority of Singapore (PSA), Singapore Telecoms, and the Civil Aviation Authority of Singapore (CAAS), were lead users of IT to boost their productivity and global competitiveness.

The diffusion of IT in the private sector was also very rapid. Combined spending on IT hardware, software, and services increased from S$209 million in 1982 to S$1.8 billion in 1991, an average growth rate of 27 percent per annum. In terms of increased computer usage by business firms, the proportion of firms using computers rose from 13 percent in 1982 to 84 percent in 1992. In terms of specific sectors, the highest penetration of computerization occurred in the financial services sector (92 percent), followed by trade (87 percent), transport (86 percent), and manufacturing (82 percent). By 1992, there was one computer terminal or personal computer for every three employees in the workplace in Singapore.

The rapid diffusion of IT adoption was accompanied by a rapid increase in the number of IT professionals. From 1980 to 1988, the number of IT professionals increased tenfold, from an estimated 850 to 8,300. The figure has increased to about 10,300 in 1991. This rapid diffusion of computer installations among organizations and the rapid growth in IT professionals also gave rise to the need for IT managers. In a survey of managers in 1980 by SIM, computer managers accounted for less than 1.8 percent of all senior and middle managers enumerated. The proportion has undoubtedly increased significantly since that time.

While engineering education at the university and polytechnic level continued its steady expansion in this phase, it was during this period that the tertiary institutions began to expand significantly their management training programs at the polytechnic diploma, undergraduate degree, and postgraduate levels. In particular, the Faculty of Business Administration at the National University of Singapore significantly expanded its part-time B.A. and M.B.A. programs to cater to the rising demand for more advanced training in business management. However, the training model remains one that emphasizes compartmentalization along strict disciplinary lines—engineering faculties training engineers, and the business faculty training "generic" management graduates and M.B.A.'s who are supposed to be able to fit into any organization and industry.

To cater to the rising demand for IT professionals, the Department of Information and Computer Science at NUS also rapidly expanded its enrollment, while the Institute of Systems Science (ISS) conducted a wide range of short training programs and longer professional-degree programs in the IT field. Although there was some attempt to introduce technology management content into these programs, the focus remained largely technical, as it did in the various programs to train manufacturing technologists such as the Singapore-French Institute, Grumman Institute for Manufacturing Technology (GIMT), and the Automation and Application Centre (AAC).

The role of private sector employers in MOT skills formation also grew in importance in this phase. An increasing number of firms, especially the larger enterprises and government agencies, sponsored their staffs' participation in formal management training programs. Short executive programs of one to two weeks' duration also became popular. Moreover, some of the more innovative companies such as Motorola provided significant in-house management development programs through specialized in-house training units (e.g., Motorola University).

Future Technology Strategy and Demand for MOT Skills

For the 1990s, as Singapore seeks to embark on the "Next Lap" to catch up with the industrialized nations under a new Strategic Economic Plan (SEP), there will need to be a strategic shift from technological development with an emphasis on *absorption* (or utilization) development focused on technological *creation* (innovation). Using the analytic framework formulated in chapter 1 of this book, we can see this shift as one that involves moving the center of gravity of Singapore's firm-level technological capability (FTC) toward a higher level in general, and more specifically toward a higher product technology capability in particular. This shift has three components: first, encouraging more MNCs to locate their product-technology-development activities in Singapore; second, increasing public sector involvement in R&D and technology out-transfer activities; and third, encouraging local firms to move into product-technology-development activities, either on their own or in collaboration with foreign MNCs or the public sector, or among themselves. Complementing this overall strategic thrust is the necessary concomitant expansion of local FTC along the other capability dimensions toward the high-level end. In particular, there is a need for local firms to improve their technology marketing capabilities, and more important, their strategic management capabilities.

Thus, while the trend toward increasing adoption of advanced manufacturing technologies (AMTs) and intensification of IT adoption is expected to continue in the 1990s, the biggest change in MOT skills requirements during this new phase is expected to come from the National Technology Plan (NTP) push to promote R&D and related innovative activities, such as industrial design and technology commercialization. Until very recently, the number of R&D management-related positions was very small due to the relatively low level of R&D activities in Singapore. In 1978, R&D expenditure as a percentage of GDP was only 0.2 percent, while the number of research scientists and engineers stood at only 818, or 8 per 10,000 workers in the labor force. Corresponding to this low level of R&D activities, a 1980 Singapore Institute of Management (SIM) survey mentioned earlier enumerated only 57 R&D managers, which accounted for only 0.8 percent of all senior and middle management jobs enumerated. There has been a steady rise in R&D activities since the late 1970s, with R&D spending reaching 1.27 percent of GDP and the number of RSEs per 10,000 workers in the labor force reaching 39.8 in 1992. Out of an estimated total workforce of 10,600 engaged in R&D in 1992 (including 6,500 RSEs and 4,100 technicians and support staff), an estimated 1,200 (about 11 percent) were identified as R&D managers (NSTB 1993).

We can expect the demand for R&D management and related technology creation skills to increase more rapidly over the next decade under the NTP program, which lists the following objectives:

- to help develop broad-based technological and workforce capabilities in nine key technological areas;
- to help develop more specialized capabilities in a few key technologies in each area;
- to set up new, publicly funded R&D centers and institutes to undertake precompetitive R&D;
- to expand assistance and support to the private sector in its developmental and commercialization efforts;

- to develop physical and other technology infrastructure to support R&D efforts.

As part of the NTP, a S$2-billion budget allocation was made for 1991–95 for various programs to achieve the above objectives. The NTP is targeted to help Singapore raise its ratio of gross expenditure on R&D (GERD) to GNP from the current 0.9 percent to 2 percent by 1995, and the corresponding number of RSEs (research scientists and engineers) per 10,000 workers in the labor force from 28 to 40. While these targets are somewhat ambitious, they are still somewhat behind the current Taiwan RSE figures and the projected Korean figures. Some progress had already been made by the end of 1992, with R&D expenditure as a percentage of GDP rising to 1.3 percent and the number of RSEs per 10,000 workers rising to 39.

Although disaggregated statistics are not available, a significant proportion of the more senior R&D management positions are held by expatriates. Given that more than half of the 331 private firms performing R&D in 1992 were foreign MNC subsidiaries or firms with foreign-majority ownership, and given that these MNCs account for a disproportionate share of the R&D spending (72.6 percent in 1992), this situation is not surprising. Indeed, it can be said that in the 1980s to early 1990s Singapore was at a level of skills development with respect to technology creation and R&D management comparable to that of the 1960s and 1970s with respect to technology absorption or manufacturing management.

In order to meet this strategic shift in technology focus, Singapore will face various new MOT challenges in the 1990s. To provide a more detailed picture of these challenges, it is useful to differentiate between MOT challenges faced by managers of individual firms in the private sector and MOT challenges for the managers in the public sector in terms of national S&T policy formulation and implementation.

MOT Challenge at the Firm Level

Product Technology Innovation Management

As more MNCs seek to redistribute a greater part of their product-technology-development activities to Singapore, there will be an increasing need for managing the cross-border innovation interface between the parent company, where market research and basic R&D are done, and the Singapore operations, where product development/

engineering and process engineering/manufacturing support will be increasingly located (Wong 1992b). In addition, as more MNC operations in Singapore evolve into "world-product-mandate centers," the Singapore operations will take on an increasing share of the total value-chain activities from design to marketing/logistics support. At the same time, some part of manufacturing operations will be redistributed to other lower-cost ASEAN countries, with Singapore serving as the regional coordinating hub. The increasing need for such value-chain activity integration and business reconfiguration will greatly increase the complexity of management.

In addition to increasing R&D activities by MNCs, many of the high-quality local original equipment manufacturers (OEMs) that have emerged in Singapore will also seek to develop their own product technology capability and establish their own brand-name products and marketing/distribution channels. They will need to learn to combine their manufacturing skills and quality assurance know-how with new product development/applications engineering know-how to achieve new sources of competitive advantage. While the development of technical R&D skills will in general be important, an even more critical need is the development of strategic technology management skills that ensure that the interfaces between R&D, manufacturing, and marketing are properly managed within a coherent strategy that encompasses all three value-chain activities.

Management of Advanced Process Technology In-Transfer and Absorption

Besides moving into more R&D-based activities, many of the world-class MNCs now operating in Singapore are expected to further upgrade their manufacturing operations in Singapore through adoption of advanced manufacturing technologies (AMTs), such as robotics, computer-integrated manufacturing (CIM), and flexible manufacturing systems (FMS). This will require not only increasing process technology absorption and innovation capabilities but also increasingly complex process engineering management skills. Moreover, the trend toward achieving ISO 9000 quality certification among both the MNCs and their local subcontractors will increase the demand for improved operations management practices.

For nonmanufacturing businesses such as transportation/logistics operations, software development, and other business services, the rapid adoption and assimilation of new technologies, particularly information technologies, continues to be critical. More than ever before, however, many Singaporean firms will need to become innovative *leaders* rather than fast followers. This will call for a higher level of technological assessment capability, the development of long-term strategic alliances with lead technology suppliers, and the ability to manage such relationships.

Management of Process Technology Out-Transfer

As Singapore becomes an increasingly high-cost location, many of the local manufacturing firms will need to redistribute part of their manufacturing operations to overseas locations while upgrading their operations at home. They will thus need to learn how to effectively transfer their more labor-intensive manufacturing operations to lower-cost offshore plants in other less-developed countries (LDCs) and subsequently how to manage these international operations efficiently using Singapore as a regional headquarters.

Besides these "internalized" modes of technology transfer, Singaporean firms will also increasingly need to learn to provide "externalized" or packaged technology-transfer services in the form of management consultancy services, engineering design and system integration services, "turnkey" project management services, technology brokerage, and other forms of technical assistance services. How to make use of these new forms of technology transactions poses another major technology management challenge.

Public Sector Technology Management Challenge

Formulation of National Innovation System Strategies

The implementation of the NTP will involve an increased effort on the part of the government to develop and implement strategies and policies to influence the development of the national innovation system as a whole. This will demand a greater level of expertise in strategic planning, technology portfolio analysis, and strategic infrastructure-development policy making on the part of the National Science and Technology Board (NSTB) and other related government agencies.

Compared with Taiwan and Korea, Singapore has much less accumulated experience in national S&T policy, planning, and program administration.

Management of Public R&D Institutes, Funding, and Programs

As in the private sector, the public sector would need to shift its technology management focus from technology diffusion facilitation to greater involvement in technology creation. The increasing volume of publicly funded investment in R&D, either through public research institutions or the institutions of higher learning, will correspondingly raise the demand for R&D management skills, including R&D portfolio planning, R&D project evaluation, and R&D project management.

University-Industry Cooperation and R&D Transfer

As a result of the increasing R&D involvement by the public sector and tertiary educational institutions, there will also need to be more highly skilled management of the transfer of university and public R&D results to the commercialization stage through improved interfaces and linkages with private industry. In addition to improving the management of the various existing institutional mechanisms, such as the science parks, incubator schemes, venture capital schemes, technology advisory panels, and various government R&D grant schemes, new forms of collaborative innovation may need to be evolved.

Emerging Pattern of MOT Skills in the 1990s: A Summary

The demand for MOT skills is expected to increase much faster in the 1990s than in earlier decades. Moreover, the composition of MOT skills demanded is expected to shift significantly toward technology creation rather than technology absorption and to focus more on technology strategy and planning skills rather than skills at the functional level. The demand for MOT skills that is emerging in the 1990s can be expected to be characterized by

- greater emphasis on technology creation management, although technology absorption management continues to be important;

- a need for more diverse training due to increasing technological sophistication of the economy;
- increased market need for more focused, integrated MOT training rather than compartmentalized technical training on the one hand and generic management training on the other;
- an emphasis on the need to balance generic MOT training versus technology-specific MOT skills training;
- increased demand for short-duration executive education programs versus long-gestation degree programs.

Emerging Responses to MOT Skills Demanded in the 1990s

Although it is still unclear how the various sources of MOT skill formation will evolve to cope with the changing pattern of demand in the 1990s as predicted above, several emerging responses from the government, institutions of higher learning, and the private sector can be discerned.

Public Sector Initiatives

The establishment of the NSTB in 1991 as a separate government agency to spearhead S&T policy planning and program implementation signaled increasing recognition by the Singapore government of the need to manage S&T on a more systematic basis. Although most of the senior management positions were filled by managers with considerable technology management experience from other government organizations, such as the Singapore Science Council (SSC, predecessor of the NSTB), Defence Technology Group, and Singapore Institute of Standards and Industrial Research (SISIR), many of the middle-level officers were engineers with little prior training in technology management. As an institution with a short history of organizational learning, the NSTB still has far to go in terms of providing systematized and institutionalized on-the-job training in S&T policy formulation and implementation skills. However, the establishment of an International Advisory Panel (IAP) comprising prominent technology managers from world-class MNCs and leading universities overseas and other international networks will help facilitate transfer of S&T management experiences, as will the emphasis of top management on wide consultation with the leaders of high-tech industries in the private sectors.

Through the NTP, the government has explicitly recognized for the first time the need to train technology managers (NTP, 41). The NTP itself suggests that local tertiary institutions should consider offering new programs in MOT to provide this training locally. This official recognition of the importance of postgraduate training in MOT provided the impetus for subsequent local university initiatives (see below). In addition, the NSTB also offered scholarship funding for professionals from the private sector to attend postgraduate MOT programs either locally or overseas, on a part-time or full-time basis.

Besides encouraging training, the government is also intensifying recruitment of experienced technology managers from overseas, at least in the short and medium term, to overcome the shortage of senior-level technology managers in various public research institutes and government agencies involved in technology planning. Indeed, in recent years, a number of the new public research institutes established by the government (e.g., the Institute of Molecular and Cell Biology [IMCB], the Institute of Microelectronics, the Information Communication Institute of Singapore [ICIS] and the Grunman Institute of Manufacturing Technology [GIMT]) are headed by experienced technology managers recruited from overseas.

University Initiatives

In anticipation of the NTP, the Faculty of Business Administration at the National University of Singapore (NUS) established a Centre for Management of Technology (CMT) at the end of 1990 to promote research and curriculum development in MOT. (This author has "championed" the establishment of this center and has been its director since its inception.) Through the CMT, the Faculty of Engineering and the Faculty of Business Administration at NUS collaboratively initiated a new postgraduate program in management of technology (MOT M.S.). The program was launched in July 1992, less than one year after the release of the NTP. Response to the program has been enthusiastic, judging by the quality and number of applicants. The program successfully entered its third year of operation in 1994 with about sixty candidates each year, and with increasing participation by international students, including students from China, Indonesia, and Malaysia. Meanwhile, the Nanyang Technological University (NTU), the second university in Singapore, also started an M.B.A. (Management of Technology) program and an M.B.A. (Management of Information Technology) program in 1993–94, though on a smaller scale.

The popularity of the university programs suggests a significant market demand for specialized MOT training at the postgraduate level; with hindsight, it might seem that this demand could have been tapped sooner by the universities concerned. It may be that the local tertiary institutions would have responded earlier were it not for the institutional inertia of the existing model of training based on compartmentalization along traditional engineering/sciences and general business management lines, a compartmentalization that is inherited from the British model of university administration.

Private Sector Employer Initiatives

Besides initiatives from government and local universities, employers in the private sector are expected to intensify their role in MOT training in the 1990s as well, through increased willingness to sponsor staff participation in formal training programs (a large number of the MOT M.S. students are on company sponsorship), as well as sponsoring more structured in-house corporate training programs, such as the Motorola University efforts. In addition, with an increase in R&D and product innovation activities being carried out in Singapore, there will be more on-the-job learning of management skills by local research scientists and engineers involved in such activities.

In addition to increased training, the recruitment of experienced technology managers from overseas is expected to continue for some time in order to overcome the initial bottlenecks in the area of R&D managers, AMT engineering managers, managers of creative design activities, and other technical function managers needed at the senior corporate level to oversee technology strategy and planning. This will be done primarily by MNCs through seconding their own managers from the parent headquarters or affiliates, although a number of emerging local high-tech companies looking to expand their R&D activities are also beginning to recruit experienced technology managers directly from overseas or by luring them away from MNCs operating in Singapore.

Compared to subsidiaries of MNCs, which can continue to draw upon the considerable resources of the corporate group, many local high-tech firms are more constrained in terms of in-house MOT skills development capabilities, and quite a few of the CEOs of these firms still do not perceive the importance of professional technology managers. In the long run, therefore, a key challenge will be to make more local high-tech firms aware of the need to invest in upgrading managerial as well as technological skills.

Initiatives in Private Sector Training Institutions

Following the development of new MOT-related programs at the local universities, a number of more commercially oriented overseas universities have also started to offer distance-learning programs in MOT-related fields in Singapore. These include the Royal Melbourne Institute of Technology (Australia), which offers an M.S. program in management of information technology; the University of Surrey (UK), which offers an M.B.A. in information technology management; the University of New South Wales (Australia), which offers a master of business and technology (MBT) program; and the Swinburne University of Technology (Australia), which offers a master of enterprise innovation program. These programs generally do not compete with the local university programs directly, as their entry and graduation requirements tend to be lower and they are designed mainly for lower and middle technical managers with weaker academic backgrounds.

Professional bodies such as the Institute of Engineers of Singapore (IES) have also begun to express interest in offering some forms of MOT-related programs for their members. A recent effort is a master in construction management program offered jointly by NTU and the IES. Various private consulting firms are also entering the rapidly expanding markets for various functional management-training programs in areas related to MOT, such as quality management and automation management.

While the postgraduate programs of the local universities and distance-learning programs of foreign universities together begin to address comprehensively the MOT training needs of middle and lower-middle technology managers, there is currently still no specialized local program that caters to the needs of top management, to CEOs and CTOs of high-tech firms, directors of R&D labs, and vice presidents of engineering. Instead, global executive programs run by top American business schools currently fill this niche (e.g., MIT Sloan School of Management's MOT executive program, Harvard Business School's program in "Building Development Capabilities," and Berkeley-

Nomura's "Strategic Management of Innovation" program), as do executive seminars by global technology management consultancy firms such as the Arthur D. Little Stanford Research Institute, and the Strategic Decision Group. As these programs tend to draw upon the experience of firms from the advanced industrial countries and may not fit the specific needs of senior technology managers from the NIEs, there may be a potential niche for new executive development programs to be fostered by business schools in the Asian NIEs, programs based on research findings and management lessons drawn from the concrete experiences and unique circumstances of the Asia-Pacific business environment.

Prevailing Influences on MOT Education

As pointed out earlier, MOT represents an emerging field of education and managerial training in Singapore, with initiatives coming from local universities on the one hand and overseas universities and private training institutions on the other. The latter sources typically import from abroad the entire curriculum contents of their programs, with virtually no attempt at "customizing" the programs to meet the unique challenges faced by Singaporean industries. As British and Australian universities tend to be most active in distance-learning programs, while American consulting firms tend to be most aggressive in pushing executive seminars and workshops for senior managers, the influences in terms of MOT theories and practices invariably reflect these sources, with an implicit bias toward traditional profession-based engineering management practices in the Commonwealth in distance-learning programs and toward large global American corporations in executive seminars and workshops.

In the case of the MOT program at NUS, the strategic intent was to incorporate the best elements of the successful MOT programs in leading American and European business schools. However, in the process of designing the curriculum, wide consultation was made with private industries (both MNCs and local firms) as well as key government agencies, and it was clear that the prevailing MOT curriculum models of universities in the United States and Europe could not be adopted wholesale, as the mix of MOT skills requirements is somewhat different in Singapore. In particular, the MOT issues facing a newly industrializing economy (NIE) are quite different from those typical in

International Conference on the Management of Technology, 28 Feb.–2 March, Miami.

Dahlman, C. J., B. Ross-Larson, and L. E. Westphal. 1987. "Managing Technological Development: Lessons from the Newly Industrializing Countries." *World Development* 15(6): 759–75.

Evenson, Robert E., and Gustav Ranis, eds. 1990. *Science and Technology: Lessons for Development Policy*. Boulder, CO: Westview Press.

Government of Singapore. 1991. *The Next Lap*. Singapore: Times Publishing.

Grange, T. 1992. "Teaching Management of Technology in European Business Schools." In *Management of Technology III*, proceedings of the Third International Conference on the Management of Technology, Miami.

Lall, S. 1990. *Building Industrial Competitiveness in Developing Countries*. Paris: Organization for Economic Cooperation and Development.

Lall, S. 1992. "Technological Capabilities and Industrialization." *World Development* 20(2): 165–86.

Ministry of Trade and Industry. 1986. *The Singapore Economy: New Directions*. Singapore: Ministry of Trade and Industry.

National Science and Technology Board. 1991. *National Technology Plan*. Singapore: NSTB.

National Science and Technology Board. 1993. *National R&D Survey*. Singapore: NSTB.

Porter, M. 1990. *Competitive Advantage of Nations*. London: Macmillan.

Simon, D. F., ed. 1994. *The Emerging Technological Trajectory of the Pacific Rim*. Armonk, NY: M. E. Sharpe.

Singapore Department of Statistics. 1992. *1990 Population Census*.

Singapore Institute of Management. 1981. *The Managerial Workforce Survey: Singapore 1981*. Singapore: SIM.

Singapore Ministry of Labour. Various years. *Singapore Yearbook of Labour Statistics*.

Wong, Poh-Kam. 1991. *Technological Development through Subcontracting Linkages*. Tokyo: Asian Productivity Organization.

Wong, Poh-Kam. 1992a. "Development of National Technological Capabilities in Singapore: Past Strategy, Future Management Challenges." In *Integrating Technology and Management: A Tripartite Approach*, ed. Karen Minden, Poh-Kam Wong, and Dipo Alam, 17–30. Proceedings of PECC S&T Task Force Workshop, Jakarta. Singapore: PECC.

Wong, Poh-Kam. 1992b. "Managing Innovation across National Borders: The Role of Champions and Gatekeepers in Overseas Subsidiaries." Paper presented at the Conference on Managing R&D Internationally, 7–8 July, Manchester, England.

Wong, Poh-Kam. 1994. "Singapore's Technology Strategy." In *The Technological Trajectory of the Pacific Rim*, ed. D. F. Simon. Armonk, NY: M. E. Sharpe.

4

Technology Management in Hong Kong: Too Little, Too Late?

E. Yegin Chen

Background

Like many Pacific Rim economies, Hong Kong faces the challenge of finding new sources of economic vitality in the new world competitive order. Traditionally a low-cost producer, its competitive advantage is being eroded by rising wage and land-rental rates and the emergence of other developing economies in the region. Seeking to upgrade its capabilities, Hong Kong finds itself divided over both the role of technology and the role of government in promoting technology development. Nevertheless, though significant initiatives in the realm of technology and management have commenced in the last decade, the development of technology managers has not received much attention.

This chapter first briefly describes the context of technology development in Hong Kong, the issues of contention regarding technology and technology policy, and the development of the government's institutions designed to oversee technology issues.

The chapter also describes the significant initiatives undertaken by government, academia, and industry to upgrade Hong Kong's technological capabilities. Many of these efforts were started only within the last decade; they cover a range of needs in research and development,

The author wishes to acknowledge the assistance from many members of Hong Kong's government, industry, and academic sectors in the course of conducting this research, as well as research assistance from Mr. Chris Cho Yue Hon.

technology transfer, and management education and training. However, the integration of technology with management is still underemphasized. To some extent, the territory's academic institutions have focused on teaching technicians, rather than on training managers to lead Hong Kong into a new world of rapid technological and managerial innovation.

Technology's Challenge to Hong Kong's Laissez-Faire Approach

With decades of remarkable growth behind it, Hong Kong nevertheless faces an uncertain future. The most fundamental issue that the people of Hong Kong have to resolve is their role vis-à-vis mainland China after 1997. As the debate about this role rages on a daily basis at all levels in the territory, it also manifests itself less directly in questions relating to the government's role in the economy, including in science and technology (S&T) matters. While the government has traditionally adopted a colonial-style laissez-faire approach, the need to prepare for a postcolonial future fraught with uncertainty has led the government to experiment with new thinking and approaches. In particular, changes and proposals aimed at steering Hong Kong into new technology-related directions have provoked considerable debate. The split within Hong Kong over political and economic issues thus extends to a less visible one over science, technology, and management.

Traditionally, Hong Kong's government has been characterized as embracing an economic policy of minimum intervention in the belief that free markets most effectively allocate resources. The government would serve to correct market failures and to "provide the necessary infrastructure and a sound legal and administrative framework conducive to economic growth and prosperity" (Hong Kong 1992, 59). Particular areas of focus in this infrastructure include industrial land and manpower for industry, as well as services that enable industry to become more productive (ibid. 79). Decades of economic success, low taxes, sound infrastructure, no subsidies for struggling industries, free conversion of foreign exchange, and ease of establishing business are just some of the benefits brought about by the minimum intervention approach. However, despite its past success, the laissez-faire policy is increasingly being questioned, particularly in the area of technology policy.

Some observers believe that the government's emphasis has indeed shifted in recent years. Instead of minimum intervention, the prevailing government function seems to be one of providing the "maximum possible support, short of intervention" (Barma 1992). This support includes government initiatives to monitor general world technological developments, track specific technology sectors, develop a scheme to provide funding for private sector R&D projects, and build an incubator for technology-oriented new ventures. Such efforts would have been unimaginable a decade ago; even today they invite controversy, because they are seen as either too much or too little government action.

Proponents of the traditional laissez-faire approach, including those from the academic and industrial sectors, argue that laissez-faire policies have proven successful in the past. Why change a winning formula? Further, they argue that the Hong Kong government should not take a technology leadership role because of several unique features of the territory that make changing such policies unwise.

First, government involvement in technology efforts cannot be justified in terms of national defense. Historical reliance upon the United Kingdom for defense has freed Hong Kong from the need to acquire and maintain defense-related technologies. Second, the limited resources of a territory of fewer than six million people not only cannot provide the critical mass needed for large-scale R&D endeavors but leave scant margin for error. Unlike large industrialized countries with an ample supply of resources and manpower, Hong Kong can undertake only a very few projects to develop S&T. It should thus be risk-averse, it is argued, as it cannot have a portfolio of projects to diversify the risks inherent in technology initiatives. In addition, continue the advocates of laissez-faire, Hong Kong's future lies in its success as a financial service center. According to this view, the territory should further its reputation as a banking center with excellent communications and quality of life; it should become the financial center underwriting China's economic growth.

Nevertheless, as one official notes, the Hong Kong government, like the stereotypical Hong Kong native, is pragmatic, rather than dogmatic. The government, industrial, and academic sectors all seek to promote continued economic prosperity. And in recent years, many members of the three sectors have come to conclude that continued prosperity requires a more pragmatic government approach toward technology development.

The community leaders pushing for a technology vision and strategy believe that Hong Kong should move to the next stage of competitive vitality by developing and maintaining an indigenous base in science, technology, and management skills in selected niches of the technological spectrum. Such a base would enable Hong Kong firms to produce goods and services with high value-added content and thereby continue playing a key role in the high-growth markets of the 1990s and beyond.

The advocates of a more active, coordinated approach toward S&T policy note that precisely because Hong Kong has been successful in the past, it needs to move to a more advanced economy. The policies that worked yesterday may not work tomorrow. In particular, the emergence of other low-cost competitors and the advent of technology-based competition with numerous risks and extended product development times requires a long-term view and active policy.

If in fact Hong Kong is destined to remain a financial center, technology would still play a vital role. Even if the territory could remain a financial service center without some manufacturing base, Hong Kong would still need to upgrade its technological capability. In a financial world increasingly dominated by sophisticated information and communications systems, the ability to apply technology to market needs will play an increasingly vital role in determining living standards. Furthermore, with the economic and financial liberalization of China, it is just a matter of time before other competitors, such as Shanghai, arise to challenge Hong Kong for the role of financier and information center. Indeed, some companies have already relocated their "backroom" information processing functions from Hong Kong to lower-cost locales such as Shenzhen; only Hong Kong's continued ability to boost productivity—with the emphasis now on information systems—can ensure the territory's prosperity. Investments in employees, management, information technologies, and communications technologies would still be needed in Hong Kong.

Regardless of the vision for Hong Kong's economic future, it is clear that its economic growth will necessitate continued structural changes toward the production of higher value-added goods or the provision of services, both of which require greater investments in technology. Even were it not facing numerous issues raised by its return to mainland Chinese sovereignty, the territory would still have to cope with the question of its role in the world economy in the face of the erosion of its cost advantage. Regardless of whether Hong Kong is to become a financial service center or a producer of technology-based products, its economy requires an upgrading of know-how in technology management skills. But what does *technology* mean in a traditional entrepôt like Hong Kong?

What Do Management and Technology Mean in Hong Kong?

Some of the debate in Hong Kong regarding technology policy stems from a lack of consensus in the territory about the meaning of *technology*. In general, government and others see technology as part of the industrial infrastructure supporting existing industries, while technology advocates view it as a means to bring new industries into the territory.

Twenty or thirty years ago in Hong Kong, discussion about technology tended to be in terms of process technologies needed to improve efficiency of manufacturing. As the economy moved toward the production of increasingly sophisticated goods, technology has become equated with products and designs from abroad. Since these technologies historically have been developed overseas and imported to Hong Kong through "reverse engineering," the discussion became centered around technology transfer through foreign investment, which was seen as the primary conduit for the new production techniques needed to keep the territory cost-competitive. As for the future, the territory does not have the scientific base to be a world leader in many areas. Hong Kong's expenditures on R&D are estimated at just 0.04 percent of gross domestic product ("R&D" 1992), among the lowest in the world. As a result, some observers believe that the territory will always be a technological follower, and that R&D in Hong Kong will always stand for "replicate and duplicate."

The countervailing view holds that Hong Kong can and should seek out, assess, and exploit opportunities in technology. It should build or import a scientific base in select areas. Discussing technology in terms of concrete business opportunities and products rather than as an abstract entity is the key to properly understanding technology in the territory. The two major approaches, which are not mutually exclusive, are based on finding "enabling technologies" and "niche technologies," respectively, technologies that Hong Kong can selectively develop, use, and be competitive in.

In this vein, the 1990 *Technology Road Maps* joint project (Kao and Young 1991) represented the cooperative efforts of six Hong Kong institutions (the University of Hong Kong, Chinese University, the University of Science and Technology, Hong Kong Polytechnic, City Polytechnic, and Baptist College) to identify technologically viable ways for Hong Kong to build knowledge-based industries. Building such industries from the ground up, it was realized, would prove an impossible task. Instead, the approach adopted was based on identifying areas where Hong Kong could provide "enabling technologies."

The concept of enabling technology rests on the idea that profitable opportunities often involve the integration of several components into a product. Many of the necessary components and the technologies for creating them already exist. The missing portions, which can be a small part of the total, are the "key to profitability in the so-called high-tech products" (ibid. 12). Firms—or peoples—that can identify and appropriate one of these enabling technologies, even very briefly, can prosper through short-term market power. Thus, though Hong Kong cannot by itself create all the elements needed for a technology-based product, by adding selectively to the world's existing knowledge base it can develop new products, particularly those suited to an economy with limited land and labor.

Along similar lines, the concept of being a niche exploiter in technology products results from recognition of the special abilities of Hong Kong and other Asian manufacturers. These firms have been known to switch product lines in a few months, using incremental advances in technology to exploit windows of opportunity in the market. Rather than committing exorbitant funds in R&D and capital-intensive efforts, the small, nimble firms of Hong Kong can specialize in rapidly assessing market needs, producing specifications, and production (Asia's High-Tech 1992, 64–77).

As a greater number of community leaders begin to believe in the importance of exploiting technology niches, they have sought greater government support to promote technology and management in selected areas. By and large, according to some interviews, although some gains are made, the experience leaves them frustrated. The next section examines how the government has viewed technology policy.

Government Nonpolicy toward Technology

As a whole, the government is not a proactive leader as far as efforts at improving the technology base; rather it is a reactive listener to market needs. From the discussion thus far, it should come as no surprise that Hong Kong, unlike many of its competitors, has no grand strategy in either S&T or MOT. The reluctance to take charge stems naturally from the laissez-faire approach toward the economy, and from the lack of a shared vision about the territory's future and the absence of a sizable group of officials and civil servants giving technology a voice in the policy process. This section provides a short overview of technology policy development.

Hong Kong's debate over official support for a proactive industrial approach can be traced back to at least the 1960s, when the government studied the need for a Productivity Council to keep industry competitive (Working Committee on Productivity 1964) and heard speeches in the legislature advocating a greater government role in the economy (as noted in the report of the Advisory Committee on Diversification 1979), particularly with regard to issues of productivity and competitiveness. Concerns about productivity and the government's role in maintaining economic prosperity surfaced again in the late 1970s, leading to the formation of a high-powered committee of officials, industrialists, and academics to study the issue.

The need to establish study committees on technology issues occurs often in Hong Kong, which raises the important issue of whether the government as a whole has an appreciation for or expertise in technology issues. Some observers claim that the government lacks the resident expertise to make informed judgments and to act on matters relating to S&T needs, forcing it to rely excessively on one-shot panels and external consultants. Indeed, the Advisory Committee on Diversification reported back in 1979 that "we have noticed a lack of knowledge and understanding within the Government Secretariat of matters within the field of applied science and technology" (ibid. 325). In recent years, the emigration of many civil servants, low civil service morale (Chow 1992), and a general lack of government resources devoted to S&T issues have only hampered efforts to build a body of expertise in the government. One senior government official maintains that training programs for civil servants and the use of external consultants and expert panels help ensure that S&T issues are dealt with by an adequate body of expertise. Nevertheless, even he notes that the occasional training courses do not focus on technology issues. Further, even with consultant reports, the question remains: how knowledgeable about technology issues are the officials reviewing the report?

The general trend in the territory has been a slow movement toward a realization of the importance of technology, however it may be defined. The support for technology seems to have been justified in most cases by treating that support as a part of infrastructure, much like education. As such, many technology-related programs in the past have been general initiatives that seem not to be targeted toward specific goals or industries.

Indeed, it appears that until very recently, the government apparatus was beset by indecision over the role of technology in the policy machinery. An indication is that the government has traditionally not been able to find a place for technology within its bureaucratic structure. Though it features twelve policy branches, two resource branches, and over thirty departments (Hong Kong 1992, 29), not one of them focuses on S&T matters. Instead, over the years a hodgepodge of ad hoc bodies and short-lived entities have played only an advisory role in technology policy issues. Only in the last year have steps been taken to establish a group of experts whose full-time responsibilities include monitoring technology developments, needs, and resources.

The Industry Department

The organization with responsibilities most closely linked to technology issues, the Industry Department has the mission of "improving the competitiveness of the manufacturing industries [developmental support], to promote production [infrastructural support], and to promote and facilitate inward investment in manufacturing and manufacturing support services" (Director General 1991–92, 12). Given the importance of promoting industry in Hong Kong, one might expect that such a department would have a long-standing, strong identity. Yet the Industry Department became an independent identity only in 1982, when it separated from what had been the Trade, Industry, and Customs Department. As a further indication of the Industry Department's lack of importance in the government's eyes, its 1992–93 budget of HK$105 million placed it approximately in the bottom 15 percent of all government departments in terms of funding.

Even after industry was deemed important enough to merit a separate department, there seemed to be some confusion about the department's organization and the role of S&T within it, based on a review of department documentation. When the Industry Department was originally formed, among its four divisions was a Science and Technology Division. This Science and Technology Division administered product standards and provided input on applied R&D and technology transfer matters (Hong Kong 1984, 24). However, through a 1984 reorganization of the Industry Department, the Science and Technology Division was "retitled" the Industrial Support Division. By 1988, after yet another departmental reorganization, what had been the Science and Technology Division seemed to be split into two divisions. One was responsible for issues relating to the provision of industrial land, workforce training, water energy and fuel supplies, materials, and regulations (Director General 1988–89, 14); the other was responsible for providing technical support, monitoring the various industrial support services, managing the government's industrial consultancy program, and planning new development support strategies. In short, instead of a major department focusing on S&T, as was originally intended, what remained were two divisions having partial responsibility in indirectly related areas.

A December 1991 effort at strengthening the technology function involved the establishment of a Technology Development Division within the Industry Department. Consisting of five officers, the division serves to provide secretariat support to the Industry and Technology Development Council (ITDC), to promote technology development in Hong Kong, and to administer the Applied Research and Development Scheme (Director General 1991–92, 14), which is described later in this chapter. Though certainly a step in the right direction, the Technology Development Division will likely require more resources and time to truly fulfill its objectives.

The Technology Advisory Function

Concurrent with the apparent confusion over the Industry Department and technology issues, the Hong Kong government has had several advisory boards on technology issues over the years. The boards benefit from the substantial expertise of members drawn from industry, government, and academia, but they could be more effective. Their primary shortcoming is that they can only advise, they cannot execute. In addition, it is not clear whom they advise: there are no hard lines of responsibility, no external party in the government formally charged with integrating the boards into the policy process. This section describes, in an incomplete manner, some of these advisory bodies.

From its establishment in 1983 until 1992, the Industry Development Board (IDB) advised the government on "all matters affecting the manufacturing industries" (Director General 1988–89, 16), serving as the government's main adviser on industrial policy matters. Chaired by the financial secretary (a post requiring no obvious expertise in technology issues), the IDB included representatives from trade, industry, academia, and the civil service.

However, the IDB's responsibilities overlapped with those of the Committee on Science and Technology (CST), established by the government in 1988. Specifically charged with monitoring technological developments and similarly composed of representatives from various sectors, this committee commissioned or made detailed studies in various technology areas, and it has tendered advice to the government on various aspects of information technology, biotechnology, science and technology infrastructure, and related matters (Hong Kong 1992, 80).

The era of two bodies, the Industry Development Board and the Committee on Science and Technology, offering—apparently uncoordinated—advice on technology matters ended in 1992, when the two were replaced by the new Industry and Technology Development Council (ITDC). The ITDC would act as the "government's principal adviser on industry and technology matters" (Director General 1991–92, 11). In particular, the new council focuses on "how industry should respond to technological developments so as to maintain its competitive edge" (Hong Kong 1992, 79). Formerly chaired by the financial secretary, the ITDC is now headed by an industrialist appointed by the governor. Its twenty-three other members are drawn from government, industry, commerce, finance, and higher education.

Though certainly the ITDC is a step in the right direction, its current structure suffers from the same weaknesses as those of its predecessors: It has no executive power to decide or allocate funds. It does not have a permanent professional staff: indeed, it relies upon the Industry Department for secretarial support! And it is not clear whom the ITDC advises, in the sense that it does not directly report to another government entity or official.

In summary, the government has belatedly devoted attention to S&T issues, although in a tentative manner and through agencies endowed only with advisory capacities. Further, rather than being an integral part of the government structure, technology issues are perceived to be compartmentalized into one small portion of the government, possibly limiting the effectiveness of decision making and implementation. This lack of government direction raises questions about the overall coordination of the territory's technology-oriented programs, which are described in the following sections.

Organization of MOT

Hong Kong MOT in the Past

Despite the historical lack of any concerted effort to develop technology and its management, the territory's public and private sectors had built a serviceable collection of organizations promoting MOT by 1983. Though quite spare by modern standards, the system existing in 1983 was designed primarily for a colony dependent upon trade and low-cost, labor-intensive manufacturing for its livelihood. This section describes the elements supporting Hong Kong's MOT a decade ago, and places these elements within a simple framework to allow an examination of their respective roles.

Using Figures 4.1, 4.2, and 4.3, we can examine the development of MOT capability over time. This capability is disaggregated into three components: R&D, technology transfer, and management education and training. For the purposes of this chapter, *R&D* includes what is popularly known as basic research, applied research, and development efforts. *Technology transfer* includes the processes involved in commercializing technology into marketable products or useful processes, including establishing linkages between research and market units, funding, and providing support for new or existing ventures. It generally does not refer to importing knowledge from abroad. *Management education and training* refers to formal full-time education at the graduate and postgraduate levels combined with assorted part-time programs created to upgrade the skills of workers, supervisors, and managers. The delineation of categories and placement of elements within the categories is to some extent judgmental.

Figure 4.1 presents the development from 1983 to 1993 of organizations in Hong Kong promoting R&D. Figure 4.2 covers Hong Kong's technology transfer efforts, and Figure 4.3 the management education and training initiatives. In each figure, the horizontal axis represents time; the vertical axis represents the years the various institutions became effective, with organizations established before 1983 shown as

Figure 4.1. **Research and Development Initiatives**

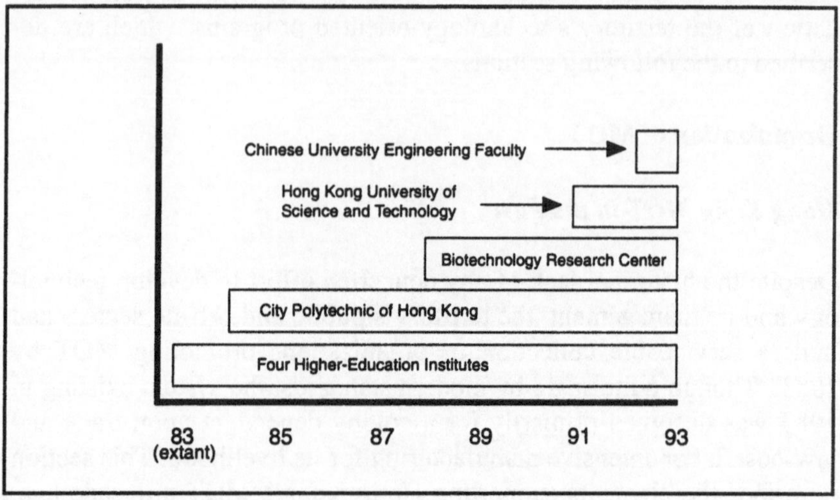

Figure 4.2. **Technology Transfer Initiatives**

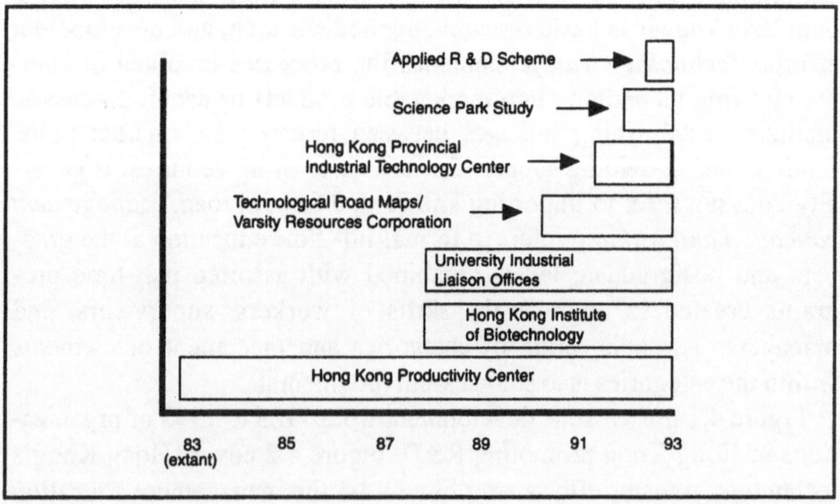

Figure 4.3. **Management Education and Training Initiatives**

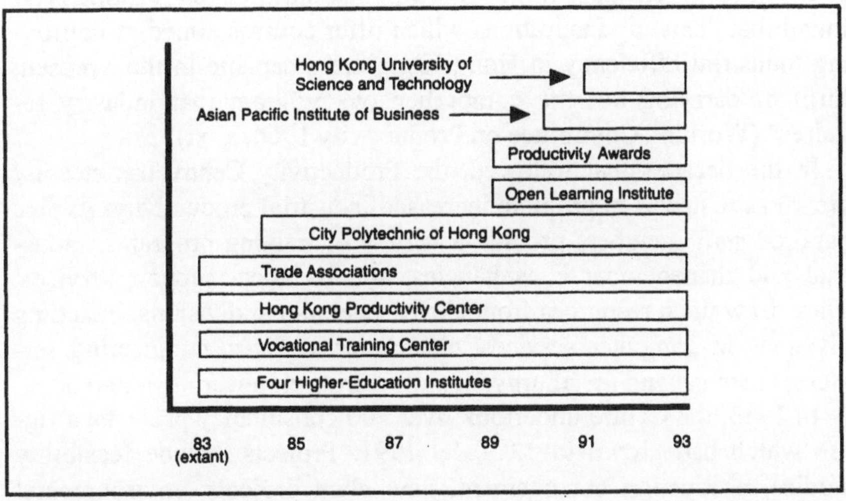

being extant in 1983. With this chart, the evolution of Hong Kong's MOT abilities can be illustrated. In particular, the development of new initiatives and organizations can be seen to have both a temporal element and a functional element.

The three figures seek to convey three main messages: that Hong Kong has embarked on a remarkable series of technology and management initiatives in the last decade, that these initiatives encompass a gamut of functions but are concentrated in technology transfer, and that upgrading the higher-education sector is a vital component of Hong Kong's strategy to upgrade MOT skills.

The following subsections describe the major organization that existed in 1983, divided into government, educational, and private sector efforts.

Government Efforts

The Hong Kong Productivity Centre

The bulk of the government's efforts to implement a policy of boosting productivity reside in the Hong Kong Productivity Centre (HKPC), established in 1967. Probably the first significant government undertaking to aid technology development, the HKPC is now considered an integral and effective aid to industry. The HKPC was started after a 1963 government-appointed Working Committee on Productivity found that "existing institutions which offer courses aimed at improving industrial efficiency in Hong Kong are incapable in their present form of carrying out the comprehensive program that industry requires" (Working Committee on Productivity 1964, x–xi).

In the decades that followed, the Productivity Centre has steadily grown as it strives to promote increased industrial productivity. Its five hundred staff members provide a variety of training programs, industrial and management consultancies, and technical support services. They draw upon resources from eleven operational divisions, including divisions in computer services, electronics services, engineering services, chemical and metallurgy, manufacturing engineering, and others.

In 1983, the Centre undertook over 200 consultancy projects, a figure which had grown to 1,050 by 1991. Projects include feasibility studies, production management, new plant projects, environmental management, quality management, product design and development, and industrial automation services.

In the realm of training, the Centre provides numerous courses and training programs applicable to a range of industries. Technology-related offerings include quality control and management, information and decision support systems, computer-aided design, ISO 9000, and specific manufacturing techniques. Management-related courses cover topics such as personnel management, marketing management, and purchasing and supply.

Vocational Training Council

Since 1982, the Vocational Training Council (VTC) provides publicly funded technical education. The VTC operates eight technical institutes, provides industrial training for eighteen industrial and service sectors, and assists employers with in-house staff training. Regular surveys produced under the auspices of the VTC determine the personnel needs of each sector, resulting in proposals for new or modified training courses.

The VTC's technical institutes provide technical training in fifteen disciplines, including clothing, computing studies, construction, design, electrical and electronic engineering, hotel management and tourism, printing, textiles, and others. In September 1991, the eight institutes offered 340 courses to 12,300 full-time and 45,500 part-time students. In addition, 18,000 working employees attended 320 short courses.

Industrial craftspeople and technicians as well as clerical and supervisory personnel can receive basic training or skills upgrading at the VTC's eighteen industrial training centers. Over 30,000 trainees attended full-time or part-time courses at these centers in 1991.

As for management education, the VTC's Management Development Centre does not provide training courses or programs but serves to promote management training. In particular, it conducts research projects and develops learning materials. One of its publications lists privately offered training courses in Hong Kong: among the over 1,500 short-term courses offered, the guide listed approximately 300 in human relations and personnel management, 150 in marketing, and 170 in communications, but only about 40 on matters relating to technology and management. Even these courses are only tangentially related to technology and tend to focus on production management and quality control (Management Development Centre 1990).

Educational Sector Efforts

Tertiary Education Sector

In 1983, the tertiary education sector featured two universities, one polytechnic, and one college recognized by the government university committee. Total full-time tertiary sector enrollment (graduate and postgraduate) was a mere 21,000; less than 3 percent of the seventeen-to-twenty age group had first-year degree places. By and large, the two universities, Hong Kong University and Chinese University, focused on training students in the arts and sciences; the training of skills applicable to industry was left to Hong Kong Polytechnic. All the institutions were essentially teaching institutions: the concepts of research and service to local industry were still not widespread on campuses.

Private Sector Efforts

Hong Kong Management Association

The association was established in 1960 to improve the effectiveness and efficiency of management in Hong Kong. It runs management training courses, provides management consultancy services, publishes a bimonthly journal on local management issues, provides library and translation services, and organizes seminars, forums, and business clubs. Training courses are offered in finance management, general management, human resources, information technology, production, and supply, among other subjects. Members join on an individual basis and include native Hong Kong managers as well as expatriates.

Other Trade Associations

While trade associations in Hong Kong are ubiquitous (there exists a sizable directory listing all the associations), many focus on functions other than training. Among the ones that do focus on training, the Hong Kong Management Association (HKMA) is the most widely known. In addition, the Chinese Manufacturers Association of Hong Kong (CMA) operates two prevocational schools providing technical education for over 2,200 students. It also awards annual scholarships to students in universities, polytechnics, and technical institutes. In general, the HKMA and trade associations offer training in management, with only a few courses in technology-related areas such as information technology and operations/supply.

Expatriates and Overseas Investment

Since its founding over a century ago, Hong Kong has received many expatriates and returning overseas Chinese and has benefited from extensive overseas investment. The territory continues to permit the free flow of both human expertise and investment capital. There are no restrictions on companies located in Hong Kong seeking to hire overseas managers and technical experts (a government guideline based on salary levels does limit the importation of manual laborers); as a result, the territory has sizable populations of Americans, British, Japanese, and representatives of other countries and many multinational enterprises. Similarly, the Hong Kong government has traditionally relied upon expatriates to fill legislative, civil service, and other official positions. A fuller description of the role played by expatriates and Chinese returning from overseas and by overseas capital in Hong Kong's development, along with initiatives to attract both, is outside the scope of this chapter. It is sufficient to note that government policies have traditionally encouraged market forces to meet the territory's demand for both.

A Decade of Building: 1984–93

In the last decade, the territory started a number of undertakings designed to increase its competence in MOT issues. The approach has been a three-pronged one, with efforts to upgrade R&D, technology transfer, and management education and training. As the following descriptions of these developments make clear, the emphasis has been on strengthening the tertiary education sector and enhancing technology transfer services.

Government Efforts

Technology-Based Industry Incubator

The Hong Kong Industrial Technology Centre (HKITC) facilitates technological innovation and the application of new technologies in Hong Kong industry. Established in 1991 by the Industry Department, the HKITC serves as an incubator for technology-based ventures, as well as a provider of technology transfer, design, development, and support services. The center's incubatees thus far include firms involved with telecommunications, pen-based computers, and artificial intelligence. It is important that a long-term view be taken in the evaluations of incubator success, since technology transfer to new business ventures requires time, as experience in the United States has shown. The center will house approximately fifty to sixty companies in its specially constructed facility.

Applied Research and Development Scheme

In February 1993 the government and the Industry Department officially unveiled the Applied Research and Development Scheme (ARDS), which was originally announced in late 1991 and which provides funding support to worthwhile applied R&D projects. ARDS seeks to encourage applied R&D in Hong Kong, particularly market-driven projects that have the potential to contribute to productivity growth, quality improvement, product innovation, or the increased competitiveness of Hong Kong industry. Projects funded would thus include product and process development efforts, with a preference for those with commercial application. ARDS's longer-term objective involves enhancing the technological capabilities and competitiveness of local industry.

With HK$200 million allocated, ARDS may fund up to half the costs of approved R&D projects in the form of equity participation or loan provision. Up to HK$10 million can be provided for projects undertaken by a single company or organization. While generally praised for its objectives, ARDS has drawn some criticism: there have been complaints that the available funds are too small to make a real difference, and that the scheme represents government meddling in the marketplace. But it is perceived as vital in meeting the needs of start-up venture financing, particularly with the Hong Kong venture capital industry's reluctance to support early stage ventures.

Science Park Study

In late 1992, a government-contracted consultant feasibility study found strong support for developing a science park in Hong Kong (Segal, Quince, Wicksteed Limited 1992). Significant numbers of technology-based corporations in both Hong Kong and around the world expressed interest in the science park concept. These firms found their activities suitable for a science park environment and also sought benefits from an enhanced image, access to technical experts, and access to China's technology and markets. The study also sketched a proposed science park concept based on a geographically distributed network of initiatives, including potential links to southern China. Several departments within the government are currently studying the report and may contract a second consultancy study to examine issues relating to location, organization, and operation of the park.

Productivity Awards

Initiated in 1989, the Governor's Award for Industry rewards and recognizes achievements in competitiveness in five categories: consumer product design, machinery and equipment design, productivity and quality, environmental performance, and export marketing. Various business associations and government bodies arrange the competitions in categories corresponding to their areas of interest. Winners enjoy enhanced prestige and the right to use the award logo for three years in advertising and other media.

Sponsored by the Hong Kong Productivity Council, the award for productivity and quality encourages, promotes, and recognizes Hong Kong firms. Firms that have demonstrated sustained improvement in productivity for at least three years have their applications assessed in three rounds of review by domestic and international experts. The following criteria are used in the awards: upgrading of products, vertical integration, value engineering, financial and productivity performance, human resources and technology management systems, and environmental management.

Academic Sector Efforts

Expansion of the Tertiary-Education Sector

The last decade has seen a sizable expansion of student enrollments in the tertiary education sector in Hong Kong, an expansion overseen and funded by the government's University and Polytechnic Grants Committee (UPGC). A decade ago, less than 5 percent of young adults aged seventeen to twenty could receive tertiary education in the territory. By 1991, the figure had grown to 18 percent. Furthermore, in 1989 the government announced plans that would boost the percentage to 25 percent by 1994–95, as well as to increase research funding. Reasons for the continuing efforts include the "demand for more skilled managers and technologists" combined with the expected emigration of many local graduates (Hong Kong University 1991, 16).

The growth has resulted from expanded student enrollment at existing institutions (Hong Kong University, Chinese University, Hong Kong Polytechnic), the establishment of new institutions (City Polytechnic, the University of Science and Technology, the Open Learning Institute), and the upgrading of existing institutions to be able to receive government funds (Baptist College and Lingnan College). Both externally oriented endeavors to integrate technology and management and these institutions' internally oriented efforts are described in the sections on technology transfer initiatives and MOT programs (see pp. 115–19).

Establishment of City Polytechnic of Hong Kong

Founded in 1984 as part of the expansion of the education sector, the territory's second polytechnic has grown to have a student enrollment (full-time, part-time, and sandwich-course) of over 12,000. City Polytechnic features faculties in business, the humanities and social sciences, and S&T. Particular areas of faculty expertise include electronics and engineering. The City Polytechnic has also aggressively developed extensive interactions with local firms, particularly on joint research and consultancies.

Opening of the Hong Kong University of Science and Technology

A new university devoted to S&T opened its doors to students in 1991. A key mission of the Hong Kong University of Science and Technology (HKUST) is to help transform the territory's economy into one that is technology- and information-based. Schools in science, engineering, and business and management offer first and advanced degrees, while faculty recruited from the world over seek to perform research relevant to Hong Kong's needs. HKUST features state-of-the-art laboratories and research equipment, and research centers in information technology and biotechnology.

The university received extensive support not only from the Hong Kong government but also from the not-for-profit Royal Hong Kong Jockey Club. This support included funds for the construction of its facilities as well as other earmarked facilities such as its Biotechnology Research Institute. Early controversy surrounding the initial construction cost overruns of HKUST seems to have subsided, as the university has apparently satisfactorily completed the second phase of its construction within budget. The university still faces the daunting challenges not only of becoming established but also of making its presence felt in Hong Kong industry and in the community through technology transfer efforts and research linkages.

Founding of the Open Learning Institute

The Open Learning Institute of Hong Kong provides distance-learning courses for adults who did not have the chance to continue beyond the secondary level and for working adults who seek further qualification or personal development. Since its establishment in 1989, the institute has grown to have an enrollment of 17,535 students; forty-four courses were delivered during 1991–92. Degree programs are offered in schools of S&T, business and administration, and arts and social sciences.

Initiatives to Promote Hong Kong Biotechnology

The Hong Kong Institute of Biotechnology (HKIB) and the Biotechnology Research Centre (BRC) were established in 1988 to develop the foundations for a biotechnology industry in Hong Kong. Generous donations from the Royal Hong Kong Jockey Club and the efforts of leading academics set up the two entities as nonprofit but financially self-sufficient organizations. Though they are not tertiary institutions in the strict sense, they are closely affiliated with universities and share common goals. They seek to promote biotechnology in the territory, in the belief that biotechnology can enhance the infrastructure of many industries, improve the quality of life, and accelerate economic development. According to the original plan, the BRC emphasizes research, while the HKIB works on development and commercialization.

The difficulty of undertaking new endeavors, combined with coordination and management issues, has precluded the rapid production of results from this biotechnology effort, leading to some skepticism within the territory. However, after a change in leadership and focus at HKIB in 1992, the institute is poised to bring commercial benefit to the territory. Rather than pursuing costly, long-term research efforts for the development of exotic drugs, HKIB emphasizes shorter-term biotechnology research and development and its application to serve the needs of existing and new industries. It will soon become an incubator for biotechnology start-up ventures, providing laboratory space, an information center, and venture funding. HKIB has already entered into several agreements with international firms and institutes that promise marketable products or processes. One of these has produced what HKIB claims is Hong Kong's first biotechnology company, a fifty-fifty joint venture between HKIB and the U.S.-based Syntex Pharmaceuticals (Wong, Fisher, and Lee 1993).

New Engineering School

As part of the upgrading of the higher-education system, the Chinese University of Hong Kong has recently initiated a Faculty of Engineering as an independent unit with a newly constructed building. While the university has traditionally offered courses in computer science, electronic engineering, and information engineering, the Engineering Faculty's formal establishment in 1991 marks its evolution from a teaching program into a teaching and research program.

Private Sector Development

Brain Gain

As previously noted, Hong Kong has benefited from the flow of technical and managerial expertise into the territory. However, a "brain drain" was sparked in the 1980s because of political uncertainty involving the territory; tens of thousands of Hong Kong residents, including many top managers and officials, emigrated.

In the 1990s, the remarkable economic growth of southern China and Hong Kong coupled with the worldwide economic slowdown has created the reverse: a "brain gain." Casual observation indicates that expatriates and returning Hong Kongese and overseas Chinese have recently been flocking to the territory, bringing with them skills and capital. Though the government has a few programs designed to attract these people (described in *Hong Kong's Training Service* 1992), they are generally perceived as having a third-order effect.

Increased Emphasis on Company Training

Company-sponsored training of employees and on-the-job training are beyond the scope of this chapter, even though they play crucial roles in developing technology managers. The tendency of Hong Kong residents to change jobs frequently creates a disincentive for firms to expend significant resources toward internal training. Evidence indicates that almost all smaller firms acquire people with needed skills by hiring them directly or by informally upgrading the skills of existing workers. Thus, the initiative has to come from individual workers seeking to advance themselves through enhanced abilities. Only the larger enterprises seem to take part in external development courses and programs. Readers interested in how Hong Kong firms upgrade the skills of their workers internally should consult *Hong Kong's Training Services* (1991) or Chan et al. (1993).

Technology Transfer Initiatives in the Academic Sector

As shown in Figure 4.2 on page 104, Hong Kong has devoted greater attention to transferring technology to the marketplace in recent years. This section highlights some noteworthy developments by universities, polytechnics, and colleges seeking to apply their expertise directly to companies in the business sector. The territory's higher-education institutions have witnessed substantial growth in three key modes of interaction with industry: sponsored research, faculty consulting, and liaison programs. The greater extent of these interactions reflects an increased commitment to serving industry needs, corporate interest in technical progress, and resource support from policy makers to provide the necessary infrastructure. Key developments in technology transfer between universities and industry are described below.

Establishment of Industrial Liaison Offices

The territory's universities and colleges have strengthened their respective capabilities to provide central focal points for initiating industry linkages and meeting the needs of all parties involved in such relationships. For example, the Chinese University of Hong Kong established the Office of Industrial and Business Development in 1988, while the University of Hong Kong founded its University Industrial Centre in 1990. These offices serve to further relationships between the university and other bodies, including local and multinational firms, and to support the commercializing of research results. Perhaps the most direct support of the commercialization of research results is the private company CPHK Enterprises Limited, founded by the City Polytechnic of Hong Kong in 1992. It serves to promote and market the polytechnic's R&D capabilities and products by providing loans to staff members starting up businesses and consultancies.

Growth of Contract Research

Hong Kong's tertiary education institutions have always emphasized research endeavors. In the last few years, the Hong Kong government has substantially increased funding for laboratory facilities and research projects. In addition, Hong Kong's universities have developed close relationships with local and multinational firms that provide financial support for research projects of mutual interest, primarily in the engineering, scientific, and business areas. Support can be through nonfinancial means as well; at Hong Kong Polytechnic, a collaborative research scheme started in 1989 provides graduate support for R&D projects under joint polytechnic and industrial supervision.

Systematization of Consulting

Though figures on the extent of faculty consulting are not readily available, circumstantial evidence indicates a significant increase recently in the provision of faculty expertise to industry. Faculty consulting efforts have become more organized and tend to be multidisciplinary in nature. To encourage direct support by academics, the institutions of higher education recently liberalized their regulation of outside practice to provide faculty members with greater financial incentives and up to one day a week to engage in such practice. Organizations such as the City Polytechnic of Hong Kong Consultancy Company were initiated to cooperate with the industrial liaison offices in marketing and facilitating consultancy projects. The Asian Pacific Institute of Business at Chinese University, which was founded in 1990, provides executive training programs in marketing and finance as well as faculty consulting services to local and multinational firms.

Initiating Changes in Relationships with Hong Kong Industry

In the last two or three years, the tertiary-education institutions have started taking a more proactive role in the technology-oriented restructuring of the territory. Among their initiatives to help establish new knowledge-based industries in Hong Kong is the *Technology Road Maps* project, an unprecedented cooperative effort involving some sixty experts from six tertiary-education institutions and from industry, with partial funding from the Industry Department. Through assessing the territory's strengths and potentials, the experts identified four technology areas that could be developed in Hong Kong: technology, biotechnology, materials technology, and environmental technology (Kao and Young 1991). This cooperative effort and the need to implement the proposals spurred the six academic institutions to create the Varsity Resources Corporation (VRC).

The VRC was incorporated in 1991 as a jointly owned company to coordinate industrial interactions among the founding institutions. Enjoying privileged access to the intellectual and facilities resources of the member institutions, the VRC has already finished at least two consultancy projects, with several more short-listed in the bidding process. It has also helped several industrial firms quickly locate academics with specific technical expertise, and it is developing ventures between Hong Kong and China.

MOT Programs

Figure 4.3 on page 105 provides an overview of the tertiary-education sector's development in Hong Kong. The previous section described the sector's extramural activities linking technology and management; this section examines its intramural programs relating to MOT.

To a large extent, the field of MOT is still in a nascent state in Hong Kong. There are no formal MOT programs at the universities, polytechnics, colleges, or private educational institutes. However, several degree and diploma programs do integrate management with particular technical areas, thus serving as a possible foundation for future developments. Below are descriptions of some of the territory's key polytechnic and university programs. (The territory's other institutions, Baptist College and Lingnan College, tend to emphasize liberal arts and social sciences.)

Hong Kong Polytechnic's Strength in Textiles and Manufacturing

Consistent with its mission, the Hong Kong Polytechnic seems to train technicians with a basic understanding of management in specific technical areas. One of its six divisions is the Division of Applied Science and Textiles. The degree and diploma students complete an integrated curriculum on both the technical and the supervisory and management aspects of textiles including courses in materials, design and pattern, clothing assembly, and dyeing technology on the one hand, and courses in clothing production management, clothing engineering, work study and ergonomics, textile evaluation, and quality control on the other. The polytechnic also offers a degree program in manufacturing engineering, which combines general technology basics (courses in electronics, materials technology, and computer-assisted manufacturing), some management courses (quality, general management, etc.), and exposure to local industry. Finally, postgraduate students already working in industry can construct individual programs based on career and employment needs. The Postgraduate Scheme in Technology combines engineering-type coursework with a dissertation based on actual employment problems to promote both technology development and technology transfer from the classroom to the workplace. The polytechnic's remaining programs are essentially engineering-oriented.

City Polytechnic's Emphasis on Construction and Manufacturing

The territory's second polytechnic has developed expertise in the area of architecture, building, and construction. Its bachelor's degree in building and master's degree in construction management feature courses in subjects such as project management combined with technical courses such as surveying and construction. A large Department of Manufacturing Engineering offers several graduate and postgraduate degrees in manufacturing engineering, engineering management, and industrial automation. The engineering management master's degree program seeks to turn engineers into managers through coursework in engineering management, communications and managerial skills, problem solving, integrative skills, operations management, project management, product management, and quantitative analysis. Though not directly linked to management, City Polytechnic's M.S. program in industrial automation seeks to help engineers make decisions regarding the development and implementation of automation technologies in manufacturing. Students learn about automation technology, the selection of automation strategies, and the planning and evaluation of automation approaches.

Offerings at Universities

The new University of Science and Technology currently features standard engineering programs, but it is working to develop programs in subjects more akin to MOT. There are scattered courses in manufacturing processes, product design, information systems, operations, and the like, but no major integration of the business/management realm with that of technology. It is planning a master's level interdisciplinary program in manufacturing engineering in the near future. The territory's oldest university, Hong Kong University, features standard industrial and manufacturing systems engineering courses, but very little integration of these with courses in management. As previously noted, Chinese University has just started an engineering program that at least initially is more oriented toward information systems.

In summary, the educational sector of Hong Kong is still far from offering an MOT program. Current programs focus on very specific technical areas (such as textiles), on the process technologies of manufacturing, or on information systems. While important, these programs may not be sufficient to help Hong Kong's technology-intensive enterprises reach their full potential. Courses in R&D management, technology strategy, entrepreneurship, new product development and innovation, technology policy and decision making, and other important areas are still lacking. The foundations do exist, however, since the educational institutions have strong management programs and strong engineering and science programs. But more efforts to link the two areas are needed.

Policy Implications

In the last decade, Hong Kong has moved away from its traditional laissez-faire policy, particularly on matters involving government support of technology. A number of initiatives involving government, industry, and academia are evidence of a more proactive approach toward sustaining the territory's economic vitality. These initiatives cover the technological spectrum, including R&D, technology transfer, and management education and training. While the recent efforts are noteworthy in the historical entrepôt of Hong Kong, two questions remain: whether the efforts are "too little, too late," and whether they can be coordinated in the absence of a central vision.

Also, despite the recent endeavors to boost the territory's capabilities in a range of technology and management areas, some gaps remain. Hong Kong needs the educational infrastructure to bridge the gap between technology and top management and policy making. At the moment, the territory's educational institutions basically provide instruction in science and engineering or in business and management, with very little integration of the two realms. There are currently no programs in MOT, technology policy, or related areas. There do exist a few programs in process-oriented manufacturing and industrial engineering, as well as programs that integrate supervisory skills with very specific technical knowledge in areas such as textiles. However, these programs are unlikely to develop enough high-level managers able to tackle the challenges of decision making on technology strategy and policy, product innovation, and managing high-technology enterprises. If managers are not just born but can be made, then Hong Kong needs concerted efforts to systematically develop technology managers. In-depth MOT programs at the postgraduate level are needed to develop managers able to run technology-based companies; these programs may combine elements of M.B.A. programs, postgraduate engineering programs, and some newly developed courses. In addition, short courses, seminars, and short-term executive programs focusing on technology management are needed to upgrade the abilities of those already in the workforce. These shorter, focused programs can assist technical specialists seeking to move into management ranks while helping seasoned executives and policy makers update their understanding of technology matters.

In summary, technology initiatives are taking root in Hong Kong. These initiatives require strong, informed leadership from government and from the industrial and academic sectors to ensure that they fulfill their potential. Furthermore, Hong Kong needs to further its commitment to its technological upgrading, particularly in efforts to integrate technology training with management training. With vision and sustained commitment, the initiatives begun in the last decade may lead Hong Kong to new sources of economic vitality; otherwise, they may indeed prove to be "too little, too late."

References

Advisory Committee on Diversification. 1979. *Report of the Advisory Committee on Diversification*. Hong Kong: Government Printer.

"Asia's High-Tech Quest." 1992. *International Business Week*, 30 November, 64–77.

Barma, T. H. 1992. "Government and Technology Development." Luncheon speech of the Hong Kong director general of industry, Science Park Conference, 3 December, Hong Kong.

Chan Lin, Candy Lau, Doris Lee, and Barbara Sit. 1993. "Training Programs in the Electronics Industry." Chinese University final year project.

Chow, King W. 1992. "The Civil Service." In *The Other Hong Kong Report: 1992*, ed. Joseph Cheng and Paul Kwong, 39–51. Hong Kong: The Chinese University Press.

Director General of Industry. 1988–89. *Departmental Report*. Hong Kong.

Director General of Industry. 1991–92. *Departmental Report*. Hong Kong.

Government Information Services. 1984. *Hong Kong 1984*. Hong Kong: Government Printer.

Government Information Services. *Hong Kong 1992*. Hong Kong: Government Printer.

Government of Hong Kong. 1993. "Applied Research and Development Scheme." Hong Kong: Government Printer.

Hong Kong University and Hong Kong Polytechnic Grants Committee. 1991. Report for the 1988–91 Triennium. Hong Kong.

Hong Kong's Training Services. 1991. Hong Kong: American Chamber of Commerce in Hong Kong.

Kao, Charles, and Kenneth Young. 1991. *Technology Road Maps for Hong Kong*. Hong Kong: Chinese University Press.

Management Development Centre of Hong Kong. 1990. *Management Course Yearbook 1990*. Hong Kong: Vocational Training Council.

Open Learning Institute of Hong Kong. 1992. *Annual Report, 1991–92*. Hong Kong.

"R&D: A Matter of Survival." 1992. *Asian Business*, January, 24–29.

Segal, Quince, Wicksteed Limited. 1992. "Hong Kong Science Park Study, Stage One: Draft Report to Hong Kong Government Industry Department—Executive Summary." Hong Kong.

Wong Kin-Ping, Stephen Fisher, and Siu-Leung Leem. 1993. "Biotechnology Can Offer Deals with Attractive Short-term Returns." *Asian Venture Capital Journal*, January, 9–13.

Working Committee on Productivity. 1964. *Report of Working Committee on Productivity*. Hong Kong: Government Printer.

5

Development of Technology Management in Taiwan

Jong-Tsong Chiang

Background

General Profile of Technological Capabilities

When the overall profile of technological capabilities is considered, Taiwan as a newly industrialized economy (NIE) belongs to the so-called late industrializer group. This group is characterized by such specific traits as recent modernization, low international competitiveness of the capital goods industries and a weak foundation in the related support services, and very high if not overwhelming dependence on generic technologies and product standards developed in the United States, Japan, and Western Europe (Ernst and O'Connor 1989, 33–75).

Taiwan's rapid industrialization, particularly after World War II, has not been based on inventions or innovations, or new products or processes. Rather, it has resorted to foreign technology and to learning through imitation and apprenticeship. Therefore, its development experience is very different from that of the "early industrializers" such as Britain, France, Germany, the United States, and recently Japan. Taiwan as a whole (and more narrowly, its leading firms) has never been able to adopt an offensive innovative strategy, which would allow it to be the first mover and to take responsibility for exploring and for educating potential customers and would thus require strong development capability as well as applied research and product design engineering. A defensive strategy has hardly been within Taiwan's capability either. This strategy, while avoiding the risks taken by the first mover, enables a country to catch up quickly with minor innovations, improvements, or product differentiation and thus also demands strong development and product engineering capabilities. Presently, Taiwan's firms can at best adopt an imitative strategy, following the leaders in more or less standardized and noncomplex technologies and relying mainly on cost advantage through protection efficiency to be competitive. (For the requirements of six innovative strategies, offensive, defensive, imitative, dependent, traditional, and opportunist, see Freeman 1982, 169–86). The personal computer (PC) industry in Taiwan, which mostly clones IBM PCs and depends on foreign-made key components, microprocessors in particular, is a good example.

Because it is also difficult for an industrialized country to maintain a technological edge in many key areas, Taiwan may have a competitive advantage in a few rather technology-intensive fields. For example, while many Western European countries finally withdrew from memory integrated circuits (ICs), Taiwan, with huge investments, has fairly strong IC production capacity. Through a costly and time-consuming learning process, some local firms have been able to fine-tune the equipment needed for production. But they still rely heavily on imports for their sophisticated equipment and cannot introduce state-of-the-art chips. When these firms entered the market, the technology and its future path were well defined, and the product designs standardized. Nevertheless, the value of Taiwan's IC as well as PC production capability should not be underestimated; it represents a precious complementary asset proving that imitators and followers need not be excluded from the international product life cycle. (Essentially, this cycle starts with a new wave of innovative products or a product class in advanced countries and then spills over to other developed countries and later to developing countries through trade and local production, eventually being replaced by new waves of innovation.) For instance, in addition to supporting many local products, Taiwan Semiconductor Manufacturing Company has used this capability to establish a strategic alliance with the Dutch Philips Company, which provides application-specific IC (ASIC) designs for Taiwan to produce on a regular basis.

Taiwan may also have a foothold in certain fundamental research realms, but generally it does not have enough complementary assets to exploit the commercial potential of newly emerging technology. Taiwan's research in high-temperature superconductivity is another good example. Partly benefiting from the connection with overseas Chinese researchers, it is among the pioneers in the world. But it lacks advanced technology in materials handling and fabrication, lacks process engineering know-how and development capability, so the research achievements are isolated.

On the whole, Taiwan's indigenous expertise (including S&T infrastructure) is weak. Lack of technologically sophisticated user-supplier links within its national boundaries confines Taiwan's firms to production of more standardized products, for which such links are not critical. And it is difficult for Taiwan to assimilate new technologies as quickly and extensively as advanced countries do, and much more difficult for her to generate innovations. In other words, Taiwan's situation is stereotypically different from that of the industrialized world. If a recently proposed conceptual model of four stages of national competitive development—factor-driven, investment-driven, innovation-driven, and wealth-driven (Porter 1990, 543–73)—is used, Taiwan's economy in general has not yet entered the innovation-driven stage.

Traditional Industrial Practices

As has been mentioned previously, Taiwan's industry has traditionally relied heavily on imported technology. The ability to absorb technology from advanced countries and then catch up with them is critical to most firms' success. Because local firms usually initially exploit imported technology, they need strong engineering capability in the field to adapt imported technology to their needs. In the meantime, most of their products have become fairly standardized, so firms reverse-engineer and try to devise ways of manufacturing them more efficiently, a strategy that depends for its success partly on comparative advantage (e.g., cheaper labor costs). For Taiwan's firms, there may still be room for process improvement, but not so much for product innovation.

Moreover, R&D activities in Taiwan's industry are normally for incremental innovation and are usually undertaken without establishing separate R&D departments or launching special R&D projects. This means that R&D is managed in an informal and implicit way.

In general, however, though technological strength has been viewed as important to competitiveness, the perspective is usually far more *technical* than *managerial*. Capability in managing technological activities is typically acquired by doing and based on working experience. If some managers and technical personnel seek managerial advice as to how to manage technology, the traditional management disciplines primarily take technology as a given or do not deem technology a critical parameter or variable in managerial activities. The point here is not that working experience and traditional management principles cannot work well; rather, it is that firms may manage technology much better if they view technological activities from a more managerial perspective and gain access to more appropriate knowledge about technology management.

A New Pressure on Industry

In the 1980s, Taiwan's industry faced several mutually reinforcing challenges. First, infringement upon foreign firms' intellectual property rights (IPR) emerged as a critical issue. In previous years, when Taiwan was considered a poor country, IPR infringement seldom brought about international litigation and financial loss because local firms had little to lose. Ironically, the recognition of Taiwan's capability to duplicate technology—which makes IPR infringement so common—helped attract orders from advanced countries and made Taiwan an important manufacturing base. In the last decade, however, Taiwan was regarded as financially very successful, as indicated by its huge foreign exchange reserve. Local industry was thus threatened seriously when advanced countries prosecuted IPR infringement.

Second, the product life cycle in many industries was shortened, forcing Taiwan's industry to imitate earlier and, as a consequence, thus be more likely to infringe upon unexpired foreign IPR (e.g., patents).

Third, a wide range of technological and innovative activities began to cluster around some relatively radical innovations, represented by information technology (IT) and, to a lesser degree, new biotechnology. In other words, a change or shift in the technological paradigm occurred. This aroused a strong need to capitalize on the tremendous opportunities offered by many new technologies. Producing novel IT products and applying IT in manufacturing processes and office services were among the main new movements.

In this new era, it became urgent for Taiwan's industry to strengthen indigenous technological capabilities in order to be a quick follower, to innovate around IPR infringement, or to imitate without such infringement.

Organization of MOT

Limitations of Transfer of Technology Management

The response of Taiwan's industry to this new trend was varied. Some firms devoted more resources to R&D; others gave up local production or moved the production base abroad. In the former case, technology management was not viewed as being significantly different from traditional management nor confined to the R&D function. In general, some difficulties have been found to be typical and substantial.

First, knowledge about R&D management, absorbed—or, more exactly, translated—almost entirely from advanced countries was not really appropriate for Taiwan's industry, which is mainly made up of small and medium-size firms. In fact, even in advanced countries, well-structured R&D management principles and techniques are more appropriate for established and large firms with relatively strong technological capabilities.

Second, lessons about innovative entrepreneurship, introduced mainly from the United States, did not fit the local conditions. Few local firms are as technologically innovative as their foreign counterparts, and a different local infrastructure usually constitutes another constraint. Taiwan's venture capital industry is a good example. It was promoted in the early 1980s, following the "American pattern" partly introduced by the Hambrecht and Quist Co. (H&Q), one of the world leaders in this industry, but the result proved unsatisfactory for several reasons. One is that there are no convenient mechanisms for public offering, and merger or acquisition is not a tradition, so the exit for venture capital is very difficult to find. Another is that upside potential is low, so downside risk should be controlled, leading to a conservative portfolio structure (with a significant portion invested in the mature business stage), which in effect reduces the distinction between venture capital and nonventure capital.

Third, there is still a lack of knowledge about how to manage such activities as technology transfer and adaptation, field engineering, reverse engineering, manufacturing efficiency, product imitation, incremental innovation, and informal R&D. In this regard the knowledge pool about local experience still seems too limited to produce more generic lessons that could be applied more widely among local firms. Therefore, the proper managerial practice for traditional technological activities is rarely well defined, and progress is largely subject to trial and error.

Overall, Taiwan's industry is not sophisticated about MOT. There is no integrated and comprehensive road map guiding local firms. In this regard, as a reference, Arthur D. Little, Inc., an American managerial and engineering consulting firm located in Boston, promulgates the so-called third generation R&D management, the characteristics of which can be condensed into three key points. First, top management seeks to create across business units, across divisions, and across the corporations a strategically balanced portfolio of R&D formulated jointly in a spirit of partnership between general managers and R&D managers. Second, R&D management seeks to respond to the needs of existing businesses and to the additional needs of the corporation while at the same time contributing to the identification and exploitation of technological opportunities in existing and new businesses. Third, general management institutes a strategic and operational partnership between R&D and other vital functions, in which R&D challenges and helps define the company's real technological needs both today and tomorrow, in addition to helping meet those needs. (Arthur D. Little 1991, 35–40).

Apparently, the paramount importance of the principles suggested by Arthur D. Little has not yet been recognized by Taiwan's firms. To move in this direction, besides education in general philosophy, more tailored education dealing with strategic and long-term issues, interfunctional issues, R&D issues, engineering/operations issues, and support-services issues is needed for top management, technical departments, and nontechnical functions.

University Education and Related Difficulties

In Taiwan, there has been little concerted effort in MOT education at the university level. The first initiative was National Cheng-Chi University's extension program for scientists and engineers, begun in 1985, but the course content is primarily devoted to traditional management. In 1991 National Chiao-Tung University established

Taiwan's first and only graduate institute of technology management offering master degrees, but it has encountered many difficulties. Several conferences on MOT have been held in the past few years by universities and the China Productivity Center; however, the contribution to local industry is questionable, because the conferences primarily call for papers, and there are few writers who are really experienced in MOT. There are several reasons for the dilemmas faced by Taiwan's universities.

First, despite the importance of MOT, educational programs are far less developed and less well structured than those for traditional management lines even in advanced countries like the United States. This has to do with two factors: the knowledge pool of industrial experience in MOT is relatively small, and MOT is still a young field of study without a strong academic base (see Chiang 1990b, 1992). For Taiwan's universities, the local knowledge pool of industrial experience is even more limited, the relevance of foreign lessons based mainly on established or innovative firms' experience is weak, and a comprehensive constellation of well-confirmed and rigorous principles and theories is not readily available even in advanced countries. Therefore, universities have great difficulty in organizing the content of educational programs in MOT.

Further, as a long tradition, student selection in Taiwan relies mainly on written examinations, not on an evaluation of working experience, academic records, study plans, and recommendations. This screening process largely excludes experienced but busy applicants from regular programs because they usually cannot compete with inexperienced applicants who have more time to prepare for entrance examinations emphasizing university-level knowledge. In reality, for inexperienced people, especially those without profound technological backgrounds, a small number of introductory courses about technology management suffice, because most of them will work in nontechnical functions and therefore need to enroll in traditional management programs.

In addition, the regular two-year, full-time program required by the government for a master's degree is too long; students working in dynamic industries and technological fields would very probably be out of date by the time they graduate. However, it is precisely these people who particularly need intensive MOT education.

Finally, there is a great shortage of experienced faculty qualified to teach experienced students. In Taiwan, few teachers in management schools have any real technological background, and few can fully understand the important characteristics of technology from a managerial point of view. Those who appear to have better potential to teach MOT for local industry are typically experienced industrial managers, many with advanced degrees in science and engineering and master's degrees in general management. But for lack of doctoral degrees in management these managers cannot usually join a management faculty without exceptional arrangements.

The difficulties of university education in MOT as discussed above are to a considerable extent rooted in Taiwan's rigid education system and cannot be resolved easily in the near future.

The Program of Japanese Management and Its Shortcomings

In contrast to university education in technology management, which is still in a state of flux, the China Productivity Center (CPC) appears to have moved forward firmly on its own track for three years. CPC is a semipublic organization that is supervised by the Economic Ministry, but which has considerable autonomy in serving industry, especially in the fields of quality management, production automation, and so on. CPC has held many one- or two-day courses in MOT for local industry. To date, CPC has primarily relied on a Japanese management organization, JMA Consultants, Inc., to provide most of the teachers and to supply the well-structured teaching materials. Students are instructed in what apparently are Japanese practices. Most of the cases studied are Japanese. Little emphasis is placed on academic or theoretical arguments.

Because Japan has performed well in changing its status from a technological follower to a technological superpower, the approach for Taiwan's industry seems to be to follow the Japanese track first and to make local adaptations later. From this point of view, CPC's strategy seems practical and desirable. It encourages straight action and avoids unfounded discussion. Though the imported Japanese management may still be more appropriate for medium-size and large companies, it is of utmost importance that Taiwan's industry learn the detailed operational basics, even in the supportive functions that the Japanese always stress. However, a recent CPC initiative may have gone too far. To further promote R&D management, CPC, sponsored by the Economic Ministry, organized a program to train consultants for industrial R&D management. About forty people, mostly mid-career managers of technical departments, were enrolled in a six-month, 360-hour series of lectures about R&D management given largely by Japanese teachers on weekends only, with seven-hour courses on each Saturday and Sunday. Briefly, this program seems to have overlooked several important points. First, R&D management, and more broadly, MOT, should be viewed from a corporate perspective, but CPC's program paid little attention to related issues in the fields of top management (e.g., strategic MOT) and management of other functions (i.e., traditional management). Second, it is doubtful that so many hours—equivalent to eight university courses, each with forty-five lecture hours per semester—are needed to teach R&D (or technology) management, because this experienced audience of managers could learn many codifiable principles and techniques by themselves. Third, there is no knowing, without any feasibility study, whether and how many local firms want to pay outside consultants to help upgrade their MOT; the dismal prospect is based on the understanding that R&D management in practice is highly technology– and firm-specific; therefore, the entry barrier to outside managerial consultants is very high.

Forums to Discuss Major Challenges Facing Industry

In addition to the educational programs of the universities and CPC, since early 1993 conferences have been held on "IPR Strategies for Semiconductors, Computers and Software," "Technology Strategies for Chemical Industry," and "Business Strategies for New Biotechnology in the Pharmaceutical Industry." They were sponsored jointly by the management and engineering schools of National Taiwan University and organized chiefly by the author of this chapter, and all received an enthusiastic response from industry.

The first conference addressed the pivotal importance of the intellectual property rights (IPR) strategies for at least two reasons, in addition to the previously mentioned generally unfavorable trend against Taiwan's industry in IPR. First, IPR (i.e., patents and copyrights) are numerous in the field of information technology, and thus it has been difficult for Taiwan's firms to avoid infringing upon these rights. Second, U.S. firms, which own the most effective IPR, have become far more protective in their IPR strategy in the past decade because of a new IPR judicial institution in the United States. This requires some elaboration.

In 1982 the Court of Appeals for the Federal Circuit (CAFC) was formed and became the only court of appeals to hear patent case decisions. The tremendous threat that this has brought to defendants has three aspects. First, the CAFC has upheld many lower-court decisions on patent validity. Second, not only the CAFC's final injunctions but its preliminary injunctions as well may require cessation of commercial activities. Third, IPR infringement, if deemed deliberate, may lead to a penalty assessment of triple the damages (for recent developments of IPR environment and practices in the United States, see Berkowitz 1993). In this new IPR environment, Taiwan—which serves as a major production base for many IT products—is in a very disadvantageous position and thus desperately requires defensive strategies in IPR.

The second conference tried to tackle the problems faced by the "traditional" chemical industry, which mainly produces commodity chemicals that sometimes cause serious environmental problems. Industrial leaders as well as national laboratory directors gathered at this conference to exchange ideas about strategies for technology innovation and transfer in order to strengthen the technical capabilities of this industry.

The third conference was held to help local firms cope with the challenges that confront them in the commercialization of new biotechnology. Although Taiwan's firms can produce some generic drugs, the local lack of experimental animals and clinical test systems seriously deprives them of opportunities in pharmaceutical production. In this conference, international strategic alliance, cooperation between research labs and industry, and IPR strategies were extensively discussed.

Future conferences under preparation include management of large-scale construction programs, international offsetting and technology transfer in high-tech procurements, and applications of IT for organizational transformation. These are recent activities in Taiwan, and their impact remains to be evaluated.

Prevailing Influence on MOT

S&T Policy

A discussion of MOT at the national level requires a review of S&T policy and a comparison of Taiwan's experience to that of advanced countries. In major industrialized countries, it was assumed that increased R&D expenditures on military, "big science," and "big technology" programs could provide positive spin-off effects. This supported the supply-side S&T policies during the early postwar era. In the 1960s, demand-side thinking began to prevail, owing to the growing concern with cost effectiveness and a better understanding of the whole process of innovation and technical change. In the meantime, many methods, such as technology assessment and R&D project evaluation, became better developed. In the 1970s, the oil crisis led to more emphasis on the integration of S&T with economic and industrial policies. (For a historical review of major industrialized countries' postwar S&T policies until 1980, see Freeman 1982, 189–206.) In the 1980s, however, there was a movement to focus on "strategic technologies" through various national programs of various forms in many countries. This new trend was in reality a response to the new "Japanese challenge," which was different from the old "American challenge" of the 1950s and 1960s. To many countries, the American challenge meant "big is beautiful" (Servan-Schreiber 1968). Because the size of a firm and the scale of R&D investment were regarded as crucial, "national champion" strategy (emphasizing firm size) and "big S&T" programs thus gained popularity in many Western countries. By contrast, the Japanese challenge is characterized by a new type of national innovation system (see Freeman 1987), which is well illustrated by Japan's Very Large Scale Integrated Circuit (VLSIC) national program in the late 1970s. As a result, many national and international technology programs emerged outside Japan in the 1980s, programs also endorsing industrial collaborative R&D.

Taiwan, as a newly industrialized economy, has no similar historical locus. Until the mid 1970s, its S&T policy was an outright supply-side one, and it had neither "big science" nor "big technology." The typical approach to S&T was to increase investment in R&D, the main argument was for long-term development, and the "last shield" was the cultivation of human resources in case there was doubt about the output. In other words, the policy was input-oriented, with little comprehension of the whole innovation process. In this era, management was not so complicated.

The energy crisis of 1977—at which time 80 percent of Taiwan's energy was based on oil, 98 percent of which was imported—together with severe international competition and rapid technological progress (especially in IT and new biotechnology) led the Taiwanese government to launch several national programs centered on eight fields: energy, automation, information, materials, biotechnology, electro-optics, hepatitis B control, and food. This national thrust in effect reoriented Taiwan's S&T policy from the supply-side rationale to a perspective emphasizing a demand-side approach and the value of output. It also meant that the government, targeting strategic technologies, began to adopt technology policy to support many related industries. Before then, the policy had been only to target some industries, that is, to pick winners and help losers.

However, owing to industry's overall weak technical capability, the government had to take major responsibility and could not just support industrial R&D at arm's length. In other words, national programs in Taiwan have a different picture from those in the industrialized countries, and MOT at the national level thus became a critical challenge.

Policy Implications

Taiwan's experiences in these national technology programs suggest many lessons (see Chiang 1990a for a more detailed discussion). One of the most important is that to launch national programs, three conditions at the national level seem crucial: there must be a national leader to acquire substantial resources and to integrate the efforts of various departments; there must be an institutional adjustment to facilitate nationwide coordination; and an overall framework must be provided for the upgrading of national infrastructure and general capabilities.

In order to strengthen still inexperienced decision making, the organization of a foreign advisory group at the national level is desirable. These foreign advisers can contribute at the program and institutional levels.

Because local industry lacks R&D capabilities, public applied-R&D organizations have to play a pivotal role in national programs. Only as industry gradually increases its own technical competence will the R&D institutes be able to adopt a more indirect role and emphasize more generic R&D, which in effect means a more "offensive" national innovative strategy.

Nevertheless, without a strong upstream research community, downstream industry, and complementary technologies, many applied-

R&D organizations are isolated enclaves. In other words, the approach of whole innovation process at the national level would expose the limitations and weak links in the national innovation system.

In selecting R&D topics, an imitative country like Taiwan could benefit greatly from technological convergence and standardization in advanced countries, because these would establish clear focuses for concerted efforts. To serve industry, R&D has to meet diverse needs and adapt delicately to application contexts, so decision making becomes much more complex. Moreover, "product-function-centered" R&D is more risky, though not necessarily more difficult, than "process-centered" R&D.

Owing to the central role played by public R&D institutes, technology transfer from them to industry constitutes a big challenge. When the technology being transferred is highly complex and dynamic, the close, long-term cooperation between donor and receiver is of utmost importance. The donor must make a deep commitment, sometimes including permanent personnel transfer, and the receiver's demand for a continuous infusion of new generation technology must be met if it is unable to reduce its dependency. In this regard, the formal organizational boundaries between both sides impose a constant uncertain, albeit not always detrimental, factor.

When process technology is huge in investment scale but the product design needs to be kept entrepreneurially active, Taiwan's experience shows that the possibility for a special pattern of "strategic alliance" comes into being. That is, the government invests in process technology while the private sector, including foreign firms, takes care of product design. This national strategy requires public organizations to be willing to cooperate with and even nurture a multitude of private counterparts.

In technical service to industry, applied-R&D organizations usually cannot perform satisfactorily because of the combined lack of interest and lack of industrial field experience. Therefore, the establishment of independent technical service task forces led by experienced industrial technologists is advisable. Moreover, the potential benefit resulting from the cooperation between R&D, technical service, and managerial service functions should not be overlooked, because they may be mutually supportive in the following way: the technical service units can provide industrial field information for R&D labs and can help the latter diffuse technology to industry; the R&D labs can "refresh" and "recharge" technical service units; and the managerial service complements technical service in related management improvement. To enhance cooperation, these functions could be placed within the same organizational boundaries, under the same leadership, or at the same locations.

Because national programs are endorsed by government, executed mainly by R&D institutes, and generally fostered by grand S&T development plans, a great push for performance would compel these participants to improve themselves. In Taiwan, evidence points to the upgrading of S&T advisory and project evaluation functions in administrative systems, strategic repositioning of R&D institutes, and cross-fertilization of general plans and specific programs as the possible effect of these national programs. In other words, such programs could alternatively exert far-reaching influence on the national development context and infrastructure.

Unlike those in industry, the managerial challenges in the above-mentioned areas are very difficult to articulate, because a wide variety of complicated technological, economic, political, administrative, and even legal issues are involved. In Taiwan, National Taiwan University has begun to tackle this topic. A forum sponsored by the management and engineering colleges of National Taiwan University was organized in March 1992 mainly by the author of this chapter. This forum has so far held forty-five half-day conferences with speakers and commentators invited from government, industry, and academia. But the results are still very limited, and more in-depth studies are needed.

Concluding Remarks

This chapter tries to present an overall picture of Taiwan's experiences in the development of MOT. Without inclusion of detailed information and statistics, some points may appear oversimplified and subjective. However, the intent has been to roughly delineate a pattern that can be used as a guide to understanding many countries' development, so that strategic thinking about how to avoid some pitfalls and get through typical bottlenecks may follow.

Clearly there is a wide gap between demand and supply in technology management. Nevertheless, Taiwan cannot rely, without great adaptation, on theories of R&D management, technical entrepreneurship, IPR practices, and so on that are imported from advanced countries. In education and training, Taiwan's experiences suggest that the educational system must have the flexibility to enroll experienced students and to recruit experienced teachers. Another lesson is that action should be taken carefully despite the great need for MOT. In this regard, an intensive course to train R&D management consultants for local industry may have been inappropriate.

Many forums and conferences have been organized at the sectoral and national levels to cope with major managerial challenges. To date, the results are limited because the issues are very complex. Therefore long-term investment is needed to develop appropriate MOT training in Taiwan.

References

Arthur D. Little, Inc. 1991. *Third Generation R&D: Managing the Link to Corporate Strategy*. Boston: Harvard Business School Press.

Berkowitz, L. 1993. "Getting the Most from Your Patents." *Research-Technology Management,* March–April, 26–31.

Chiang, J. T. 1990a. "Management of National Technology Programs in a Newly Industrialized Country—Taiwan." *Technovation* 10(8): 531–54.

Chiang, J. T. 1990b. "The Research Strategy in Management of Technological Innovation." *Technological Forecasting and Social Change* 37: 267–73.

Chiang, J. T. 1992. "Technology and Management Integration: A Challenge to University." Paper presented at the Pacific Economic Cooperation Council (PECC) S&T Task Force Workshop on "Integrating Technology and Management: A Tripartite Approach," 9–10 November, Jakarta.

Ernst, D., and D. O'Connor. 1989. *Technology and Global Competition: The Challenge for Newly Industrializing Economies*. Paris: Organization for Economic Cooperation and Development.

Freeman, C. 1982. *The Economics of Industrial Innovation*. Cambridge, MA: MIT Press.

Freeman, C. 1987. *Technology Policy and Economic Performance: Lessons from Japan*. London: Pinter.

Porter, M. E. 1990. *The Competitive Advantage of Nations*. New York: Free Press.

Servan-Schreiber, J. J. 1968. *The American Challenge*. London: Hamilton.

6

Management of Technology in Korea

Ji-Soo Kim

Background

Korea was one of the poorest countries in the world after World War II. Following the civil war of 1950–53, Korea's per capita GNP was much lower than its neighbors'; many developing countries, including the Philippines and Pakistan, were far better off than Korea. But in the past three decades, Korea has become one of several successful model countries in economic development.

Overview of National S&T Policy

Since 1962, during the course of Korea's Five-Year Economic Development Plans, which emphasized rapid industrialization, the development of both technology and human resources was increasingly recognized as a key factor in maintaining growth and enhancing the international competitiveness of Korean products. Korea adopted scientific and technological development as a major policy goal and established the Ministry of Science and Technology (MOST) in 1967 as the central agency for its administration. The Planning and Coordination Office, Research Coordination Office, Program Development and Promotion Bureau, and Technical Cooperation Bureau are the primary offices of MOST, with the Office of Atomic Energy, Central Meteorological Office, and Office of Geological Survey being subsidiary organizations. In the next dozen years the ministry added more subsidiary organizations, such as the National Science Museum, the National Astronomical Observatory (which later merged with the existing Institute of Space Science and Astronomy), and Daeduk Science Town Administration Office.

To effectively achieve development of science and technology (S&T) parallel with industrial development, in 1981 MOST decided to reorganize the sixteen government-supported research institutes under the control of various government agencies and merge them into ten institutes under the administrative control of the ministry. In 1985 MOST took a decisive step, adopting a policy-oriented organization in anticipation of Korea's becoming a more advanced society in the twenty-first century.

Major Functions

The Planning and Management Office is responsible for the integration and management of the policies and programs of the ministry and the formulation and implementation of budgets, evaluation of major projects, and improvement of administration. The R&D Policy and Coordination Office is in charge of the establishment of basic plans for research and development and of the overall coordination of national R&D projects and the management of R&D projects at government-supported research institutes. In the Technology Policy and Development Office, the director general for policy planning handles overall planning and policy making for the promotion of science and technology and the establishment of a long-term S&T plan. The director general for technology policy is responsible for promoting the development of industrial technology and engineering services and for the promotion of the systems and information industry. The director general for manpower policy oversees personnel planning, policy making, and training for science and technology, and the promotion and development of a repatriation program for Korean scientists and engineers living abroad. And the director general for technology cooperation is responsible for planning and policy making for the promotion of technology transfer and for the promotion and development of S&T cooperation with foreign countries and international organizations. The Atomic Energy Bureau's function is the establishment of nuclear R&D policy for peaceful uses of nuclear energy and the establishment and management of an international nuclear cooperation policy. The Nuclear Safety Assessment Office has the task of authorizing the production, construction, ownership, management, and operation of nuclear reactors and nuclear fuel cycle facilities and authorizing the acquisition, production, import, export, ownership, and management of nuclear materials.

S&T and the Economy

Since the early 1960s, when the First Five-Year Economic Development Plan was initiated, the Korean economy has achieved rapid growth through the successful implementation of the six Economic Development Plans completed to date. Not only has the Korean economy expanded substantially, but its economic structure has been modernized and improved. The Korean economy has become mature and now faces new challenges in order to reach the advanced stage. In the early 1960s, emphasis was placed on import substitution for such key industries as cement, oil refining, and fertilizers on the one hand and social overhead capital for roads, railways, and electrification on the other.

Since the mid 1960s, Korea has put into effect a policy of export-oriented industrialization with the aim of increasing exports of labor-intensive light industrial products to the world market. Development of the textile, plywood, and footwear industries was strongly emphasized, which turned out to be very successful, partly due to the global trend toward free trade. The industrialization policy centering on labor-intensive light industries began to reach its limitations due to wage increases, inadequate technological accumulation, and the recent participation of less developed countries in the world market.

From the mid 1970s, therefore, efforts have been directed toward strengthening the industrial structure and providing for self-sustained growth through full-scale development of heavy and chemical industries. In the midst of the quantitative expansion of the economy, however, special attention has been paid to such qualitative issues as social and economic equity, beginning with the Fourth Economic Development Plan.

After the successful quantitative expansion and qualitative improvement beginning in the early 1960s, the Korean economy began to decelerate, mainly due to the second oil crisis and the consequent worldwide economic recession. In 1980 in particular, it underwent the hardships of negative growth because of the additional burden of an unprecedented bad harvest. Though indications of a slow recovery appeared from 1981 to 1984, the Korean economy as a whole still suffered from the recession.

In 1982, Korea launched the fifth in its series of Five-Year Economic Development Plans, which have provided the primary guidelines for Korea's economic development. The fifth plan was formulated with three main goals in mind: economic stability (especially with regard to prices), improved efficiency, and better balance in economic activities. These three goals were seen as the best means by which to recover from the economic setback of 1980 and 1981 and lay a foundation for stable long-term growth.

The Korean government pushed forward with the Sixth Five-Year Economic and Social Development Plan, which covered 1988–91. The plan was aimed at achieving a well-balanced society in the twenty-first century, with emphasis on increasing efficiency and social balance and on attaining stability, which has been a goal from the beginning of the initiative. Particularly since 1988, the focus of Korea's economic development has been autonomy, trade, and international cooperation, in order to achieve harmony and balance among sectors, thus enabling the general populace to enjoy a better life.

Recently Korea has faced economic difficulties arising from increased labor costs, appreciation of the Korean won, the opening of its domestic market, and other such factors. This can be traced to the fact that the Korean economy failed to properly adjust its management strategy to the changing economic and social environment. To overcome these difficulties, the development of science and technology is considered crucial. Here the importance of the management of technology (MOT) increases and is spotlighted by many businesspeople, research scientists, policy makers, and policy implementers.

Although to a large extent the Korean economy was chiefly agricultural until the early 1960s, Korea has now become an industrializing state through the success of its industrialization strategy. Table 6.1 summarizes Korea's economic development from 1962 to 1991.

Between 1961 and 1987 the agriculture, forestry, and fisheries sector's share of GDP fell from 38.7 to 11.4 percent, while the mining and manufacturing sector's share increased considerably from 15.4 to 31.5 percent, with social overhead capital also expanding substantially to support this industrialization. In 1990 the agriculture, forestry, and fisheries sector fell to 9.1 percent, the mining and manufacturing sector

Table 6.1

Major Economic Indicators from 1962 to 1991

	Unit	1962	1976	1987	1988	1991
Population	1,000 persons	26,513	35,849	42,082	42,593	44,094
GNP	Current prices in U.S.$ billion	2.3	27.4	128.4	169.2	240.0
GNP growth rate	Percent	2.2	15.1	12.8	12.2	7.5
Per capita GNP	Current prices in U.S.$	87	765	3,098	4,040	5,500
Domestic savings rate	Percent	3.3	23.1	36.3	37.3	33.5
Investment rate	Percent	12.7	25.5	29.4	29.9	31.6
Current account	Current prices in U.S.$ million	–55	–313	9,854	14,161	6,000
Trade balance	Current prices in U.S.$ million	–335	–590	7,659	11,445	6,000
Exports	Current prices in U.S.$ million	55	7,815	46,243	59,648	79,500
Imports	Current prices in U.S.$ million	390	8,405	38,584	48,203	76,500
Invisible trade balance	Current prices in U.S.$ million	43	–72	977	1,267	–500
Net transfer	Current prices in U.S.$ million	237	349	1,217	1,448	500
Foreign debt	Current prices in U.S.$ million	166	7,479	35,600	31,200	23,000
Overseas debt	Current prices in U.S.$ million			13,200	23,900	26,500
Unemployment rate	Percent	8.2	3.9	3.1	2.5	3.5

Source: Government of Korea, Economic Planning Board, 1992. *Major Economic Indicators.*

Table 6.2

Industrial Structure as Percentage of GDP

Economic Field	Year 1961	1971	1981	1987	1988	1990
Agriculture, forestry, and fisheries	38.7	27.0	18.0	10.5	10.8	9.1
Mining and manufacturing	15.4	22.3	30.9	33.0	32.4	29.6
(Manufacturing)	(13.5)	(21.1)	(29.5)	(32.5)	(31.6)	*
Social overhead capital	9.1	13.3	16.5	19.7	20.4	61.3
Other service	36.8	37.3	34.6	36.7	36.4	*

Source: Government of Korea, Economic Planning Board. 1992. *Major Economic Indicators.*

*Figures not available.

dropped to 29.6 percent, and the service sector comprised the remaining 61.3 percent of GDP. See Table 6.2 for a comparison of the percentages of GDP for the several sectors.

Since the 1970s, large-scale investment projects have been undertaken in the steel, shipbuilding, petrochemical, and machinery industries in order to deepen Korea's industrial structure. As a result, the share of manufacturing value-added contributed by light industry declined, while the share contributed by the machinery, primary metals, and chemical industries increased.

Science and Technology Development

Science and technology has been an important instrument in effecting national development policy since the early 1960s, and as a consequence, S&T policy has been coordinated with Korea's economic policy. The main feature of the S&T strategy to support industry was to lay a foundation for industrialization through the development of import-substitute industries, the expansion of light industries, and support for producer goods industries.

The science and technology strategy was intended to strengthen scientific and technical education, to build up the technological infrastructure, and to promote foreign technology imports. During this period, the Ministry of Science and Technology was established to undertake S&T development. The Korea Institute of Science and Technology (KIST), a comprehensive industrial technology research institute, was established in 1966. The following year the Science and Technology Promotion Law was enacted.

In the second stage, during the 1970s, the science and technology strategy was aimed at strengthening technical and engineering education in heavy and chemical industries, improving institutional mechanisms for adapting imported technology, and promoting research to meet industrial needs. This strategy was designed to support the government's efforts to expand the heavy and chemical industries.

In line with these strategies, government-supported specialized research institutes were established in such fields as machinery, shipbuilding, marine science, electronics, and electricity. The Technology Development Promotion Law and the Engineering Services Promotion Law were also enacted. In the course of industrialization, the manufacturing sector has expanded substantially and has contributed extensively to Korea's economic growth. Since the mid 1970s, the industrial structure has deepened remarkably. Important tasks facing the country in the 1990s are the maintenance of the current growth rate and particularly the further enhancement of manufacturing efficiency. In this context, in the third stage, during the 1980s, Korea's industrial policy was directed at transforming the industrial structure into one of comparative advantage, expanding technology-intensive industries such as machinery and electronics, and encouraging development of the technical workforce and enhancement of productivity.

To this end, in the science and technology fields, Korea has continuously supported the development and acquisition of high-level scientists and engineers with an extensive policy that includes reinforcement of graduate school education, expansion of overseas training programs, and the repatriation of Korean experts living abroad. To promote the productivity of research and development, the implementation of select large-scale R&D projects has fostered joint efforts among R&D institutes, universities, and industry; at the same time, joint research programs and joint ventures with other countries have been promoted. By strengthening indigenous R&D capabilities and introducing advanced technology from abroad, Korea has localized the key industrial technologies. Figure 6.1 summarizes the development of Korea's S&T strategy from the 1960s to the present.

Korea, recognizing the economic importance of science and technology, has formulated and implemented S&T development policies in

Figure 6.1

Outline of S&T Development Strategy

Period	Industrialization	Science and Technology
1960s	• Develop import substitute industries • Expand export-oriented light industries • Support producer goods industries	• Strengthen S&T education • Deepen scientific and technological infrastructure • Promote foreign technology imports
1970s	• Expand heavy and chemical industries • Shift emphasis from capital imports to technology imports • Strengthen export-oriented industry competitiveness	• Expand technical training • Improve institutional mechanisms for adapting • Promote research applicable to industrial needs
1980s	• Transform industrial structure to one of comparative advantage • Expand technology-intensive industry • Encourage personnel development and improve productivity of industries	• Develop and acquire top-level scientists and engineers • Perform national R&D projects efficiently • Promote industries' technology development
1990s	• Adjust industrial structure and improve productivity • Promote balanced regional development	• Develop basic science, high technology, and social welfare technology • Expand S&T resources and promote efficiency

an organized manner. Since the launching of the First Five-Year Economic Development Plan in 1962, S&T development has been deliberately and systematically introduced into integrated development planning to provide Korea with the means for reaching its overall development targets. Particularly, the country has emphasized the promotion of science and technology with the intention of benefiting the general public.

Korean S&T policy emphasizes creation rather than imitation, development of selected high-tech areas for the twenty-first century, improved science programs for students, and the securing of stable financial resources for R&D investment. For an effective science policy mechanism and institutional infrastructure, Korea's policy on S&T development concentrates on the following components.

First, Korea continuously pursues an S&T policy that allows science and technology to play a leading role in promoting socioeconomic development. For this, available resources are allocated for the promotion of science in a preferential manner, attracting high-caliber human resources to the S&T field by enhancing the social environment for science and technology. Policies and systems that are liable to be easily diversified are introduced in order to secure consistency and comprehensiveness from the viewpoint of technological innovation. In addition, the Presidential Science and Technology Advisory Committee, consisting of representatives and experts from industrial, academic, and scientific fields, makes recommendations to the president about the most desirable S&T policies.

Second, Korea has placed special emphasis on securing and nurturing cadres of creative scientists and high-caliber technological personnel in order to meet the rapidly increasing demand for R&D in both the public and private sectors. As of 1990, the total number of qualified scientists and engineers stood at 70,500, indicating a ratio of 16 persons per every 10,000, showing that Korea is still short of high-caliber personnel. To alleviate this shortage, government, together with the private sector, is pursuing the realization of effective training in S&T-related fields in higher education. In accordance with the long-term forecasting of high-level S&T personnel requirements, Korea plans to reach the goal of 150,000 scientists, or 30 per 10,000 persons, by the year 2001. Among them, 15,000, or 10 percent, will be tapped as top-level scientists capable of carrying out the leading roles in their respective R&D and engineering fields. To this end, Korea tries to reinforce higher education in science and engineering, expand overseas training programs for advanced study, and increase the repatriation of Korean scientists and engineers living abroad.

Third, Korea has drastically increased its R&D investment, recording remarkable achievements since 1980. The budget for science and technology has been increased by about 15 percent annually over the past six years, while public enterprises, including telecommunications and electrical power companies, are encouraged to set aside a sizable portion of their revenues for technology development. Along with this, the private sector has also augmented its R&D investments by about 40 percent annually, inspired in large part by various government incentive programs. As a result, the total amount of R&D expenditures jumped from $577 million in 1981, or 0.9 percent of GNP, to $4.4

billion in 1990, or 1.91 percent of GNP. However, more extensive funding for research and development is required if Korea wants to upgrade its science and technology to the level of an advanced country's by the turn of the century. Therefore, it has set the goal of boosting R&D investment to 5 percent of GNP by the year 2001.

Fourth, since 1982 Korea also has undertaken national R&D projects that could not be pursued by industry alone, in order to develop key industrial technology for its industrialization. The criteria for selecting national projects include their technology intensiveness, international comparative advantage, conservation of energy and resources, growth potential, and contribution to social development. Industry-oriented projects of this nature in such areas as energy and resource development have been carried out via the joint efforts of industry and the public sector, and projects in health- and environment-related areas have been undertaken by the public sector itself. The Korean government is responding to the ever-increasing scale and sophistication of R&D in a systematic way by upscaling national R&D projects with increased investment.

Fifth, in order to effectively utilize R&D resources, which are not sufficient in absolute scale, and to push ahead with policies systematically, Korea redefined the role of government-supported R&D institutes, universities, and private enterprise in carrying out R&D activities. The R&D capability of universities and private enterprise was nominal in the 1960s, with most R&D activities being implemented by government-supported institutes. This has changed drastically, particularly in the 1980s. Private businesses have set up their own in-house R&D facilities, greatly enhancing R&D capability, while universities have also increased their R&D capacity, allocating more expenditures for R&D. In line with this trend, universities pursue basic research, while private institutes undertake R&D for commercialization. Government-supported institutes link universities and private institutes, and they are also responsible for research that is too costly and high-risk to be done privately. They also undertake R&D in commonly based technologies that have wide application.

Sixth, in order to foster the development of industrial technology, Korea provides various incentives to private enterprise, which must play the leading role in the development of industrial technology in an open-market economy. In the case of proprietary technology, the government makes indirect incentives available to the private sector to encourage healthy competition. As for generic technology, the government extends direct or indirect support for cooperation between the private and public sectors. Following these two basic principles, the government recommended that large-scale companies establish at least one research center per company, while small and medium-size companies are encouraged to organize R&D consortia in related fields; incentives include tax exemptions, special depreciation, financial grants, and the availability of long-term low-interest development loans, among others. As a result of this incentive system, Korea has witnessed a remarkable increase in the number of private research institutes—from 52 in 1980 to 831 with 37,481 researchers in July 1992. If we count branch institutes separately, the number becomes 1,348. This means an increasing need for technology management. In addition, in accordance with the new emphasis on autonomy and in order to open its markets under the free-market principle, Korea wants to minimize controls and restrictions on industry. Thus the country virtually liberalizes technology inducement, revises measures protecting the local innovator, and extensively overhauls the laws and regulations related to science and technology in order to expand the scope of autonomy.

Seventh, Korea has strengthened international technical cooperation in order to meet the growing internationalization of technology development. To better cope with the growing interdependence of the world economy, Korea has been expanding both bilateral and multilateral international cooperative activities, especially since 1980. Korea promotes the complementary exchange of researchers and technical information, joint R&D programs, and active participation in international cooperative programs. In addition the country wants to expand technical assistance to developing countries in order to share experiences and technologies for their common prosperity.

Eighth, domestically, the regional dispersion of technology development is regarded as one of the major policy directions. Korea has established industrial complexes in major regions of the country since the 1970s, actively promoting the construction of Daeduk Science Town along with new technology-based industries in the vicinity. Korea also plans to construct other specialized research parks in major industrial areas throughout the country, forming a nationwide "technobelt" where full utilization of human resources, technology, and information is possible through the networking of telecommunication systems, transportation, and utilities.

Ninth, a climate favorable to the development of science and technology has been a major policy goal. The Korean government, with the cooperation of the academic and industrial communities and of the mass media, has launched a nationwide science movement. The movement's main objectives are to create an environment in which the general public can apply scientific principles to daily living, to involve young people in science, and to instill in them an inquiring attitude and a spirit of rationality, efficiency, and creativity. In this context, MOT is emphasized in terms of efficient utilization of limited resources, that is, investment. Korea plans to launch a science movement in a new dimension designed to enhance the ability of the people to adapt to a modern industrialized society, thereby providing a strong foundation for national development.

Korea set an ambitious goal of becoming a developed country by the year 2000. For realization of this goal, it prepared a Long-Range Plan of Science and Technology Toward 2000. With limited available resources, Korea realizes the difficulty of competing against advanced countries in every field simultaneously. Areas of comparative advantage have been selected based on certain criteria, and the country will push ahead intensively in these areas.

Based on this goal of becoming a developed country, science and technology areas are categorized into the following five groups:

1. Areas which are economically feasible in the short term, such as informatics, fine chemicals, and precision machinery;
2. Areas where the possibility of medium-term success is high, including biotechnology and new materials;
3. Areas of public welfare, such as the environment, health, and welfare;
4. Oceanography and aeronautics, where the future prospects are promising in the medium and long term;
5. Basic science and engineering, providing the basis for development in all sectors of science and technology and offering the prospect of short-, medium-, and long-term success.

Five strategies are employed for the effective implementation of these five S&T groups. The first strategy is *specialization*: technological development is to be pursued in specific fields with the limited available resources that Korea has for R&D. The second strategy is *cooperation* in the development of technology and the systematization of R&D capability by constructing a cooperative research system among industry, academia, and research institutes. The third strategy is *internationalization* of R&D in order to overcome the limitations of domestic R&D capabilities. The fourth strategy is *localization*, which involves the formation of R&D networks across major regions of the country. The final strategy is *autonomy*, by which the private sector is allowed to respond to and benefit from market mechanisms.

By applying these strategies in harmony and balance, Korea hopes to accelerate its S&T innovation and realize its goal of joining the ranks of advanced countries by the turn of the century. The nation has prepared a plan for the development of high technology and industry, with priorities in strategic fields such as the information industry, mechatronics, new materials, bioengineering, fine chemicals, new energy, space technology, ocean engineering, and welfare technology.

The MOT Challenge

As we have seen, the Ministry of Science and Technology has taken the major role in Korea's technology policy making in spite of the lack of an industrial management policy. MOST is the central agency administering many of the government-supported research institutes; it has less influence on private research institutes including those of multinational corporations (MNCs). MOT policy is handled by the R&D Policy and Coordination Office of MOST. Yet many MOT decisions and activities are made by the relevant industry. The Korea Industrial Research Institute (KIRI), which is an agency for the promotion of industrial research, plays an important role in this regard, offering MOT seminars, arranging MOT tours in advanced countries, awarding prizes, and linking industries and research institutes.

Organization of MOT

Public Sector Strategies

Again, most active in the area of industrial MOT is industry itself, and KIRI. In the public sector, the concept of MOT is still at an early stage, though the Korea Advanced Institute of Science and Technology (KAIST) and the Korea Institute of Science and Technology (KIST) are leading organizations that are giving more attention to the management of Korean technology. Industry is well aware of the importance of MOT: industrial R&D expenditures as a percentage of sales have increased from 0.31 percent in 1979 to 1.74 percent in 1989. In the same period, the percentage for the manufacturing industry was 0.33 percent in 1979 and 2.01 percent in 1989. Industry made 73.9 percent of the total Korean R&D expenditures of 2,705.1 billion won in 1989, and managers of industry promote the efficient use of R&D funds. Hence, the management of technology, including evaluation techniques, the evaluation system, and managerial know-how of R&D operations, has become a hot issue in Korean industry in recent years. Thus at the request of MOST, KAIST set up its first department of management and policy in February 1992, focusing on technology management education. MOT is the subject of one of the main programs in the department. Ten graduate students were enrolled in the master's degree program in 1993. Most were from Korean industries, although three were from the military. There seems to be a growing recognition of MOT as a necessity in the public sector, notably led by MOST. The ministry even requires all managers, notably top managers who administer government-supported research institutes, to take this MOT training program. Another semipublic organization to have MOT training programs is the Korea Productivity Centre (KPC), which holds many seminars on MOT and publishes research and survey reports. KPC is a member of the Asian Productivity Organization (APO).

Technology management is not considered a major academic discipline though several universities offer MOT as one of the requisite courses in business administration in their schools of management. Since a business administration major in general is popular in Korea regardless of the degree level (i.e., B.A., M.A., or Ph.D.), almost all Korean colleges and universities offer a business administration program. Since MOT is considered part of the business administration curriculum, the academic sector delivers MOT training in one way or another. One of the most widespread and popular programs is the senior executive program offered by many business schools. Since many senior managers, particularly middle-aged or older managers, had little opportunity for training in business administration or MOT when they were younger because of the Korean War and the postwar social instability, enrollment in these executive programs and similar MOT programs is high. These programs benefit the schools both academically and financially.

Engineering schools normally provide quite technical education and training, except for industrial engineering (IE) and engineering management (EM) departments. These departments offer courses in engineering economics, organizational behavior, R&D management, cost-benefit analysis, operations research, quality control, and entrepreneurship.

Academic networks are provided domestically through the Korea Business Administration Association, Korea Operations Research and Management Science Society, Korea Institute of Industrial Engineers, and Korea Engineering Management Society, and internationally through the IEEE on engineering management, American Society for Engineering Management, International Federation of Operations Research Society, and Japan Management Academy.

Private Sector Organization for MOT Training

Industry puts significant emphasis on MOT training. As stated above, the industry share of S&T investment in 1989 was 73.9 percent of total R&D expenditures. Thus industry, particularly the Samsung company, has shown a keen interest in MOT, particularly to promote efficient utilization of money spent for R&D. Samsung was the first company to build its own research and education institute, the Samsung Advanced Institute of Technology, in a suburb of Seoul. Not only Samsung but other companies, including Lucky-Goldstar and Daewoo, have built their own education and research institutes. All regard MOT as important. These in-house colleges have been so popular that the Korean government has permitted the graduates of these programs to receive appropriate college degrees if the students have taken all the necessary classes and units. Korea Electric Power Company also has an in-house training program that emphasizes MOT, but small and medium-size industries (SMIs) cannot afford to have individual training institutes. Thus the SMIs usually train their employees with those of other companies in order to have a large enough group. Often the training is done at the training institutes of small and medium-sized industries. There are many private consultants for MOT training, and the Korea Industrial Research Institute also provides MOT training. Educational institutions like Soongsil University focus their programs mainly on issues facing SMIs. Soongsil University even has a unique graduate school for the employees of SMIs. MNCs have their own MOT programs, which they normally operate themselves or engage private consultants to run. The Korean government plays little role in private sector MOT. All the business decisions are made by the industry itself. The sector which is devoting the most attention to MOT is definitely manufacturing. As technology becomes more complex, and different technological processes and innovations interact, technology managers serve people by managing resource utilization.

Mechanisms for Interaction

Industrial managers have abundant real-world, on-site experience, while academicians have more depth in research and the theoretical aspects of MOT. These two approaches are complementary. Universities supply graduates competent to fill managerial positions. Hence industry and academia cooperate with each other in MOT. The public sector utilizes both industry and academic experts to assist in the policy-making process for MOT. Continuous contact through consulting, seminars, conferences, and contract research are the most common and effective means of MOT. Another strategy to enhance MOT is to dispatch students for education and training through diploma or certificate courses. Policy advisory board members are in many cases noted experts and scholars. Government officials are less involved in industry except through more formal and nonfinancial ways. Mechanisms for government-industry-academic cooperation have been expanded and refined. A further example of promoting interaction through S&T infrastructure is the Daeduk Science Town, which plays a role in promoting the interaction of MOT practitioners through regular consultation meetings involving the heads of the research institutes. This is not limited to the science park, but includes frequent consultation meetings held by the heads of industrial research institutes. For this, KIRI plays an important role in industrial MOT activities. Another noted example is the Pohang steel complex, which includes a research/science park. Here the industry (Pohang Iron and Steel Company [POSCO]), the research institute (Research Institute of Industrial Science and Technology [RIST]), and the education institute (Pohang Institute of Science and Technology [POSTECH]) are located in the same complex and cooperate with each other to solve both technical and managerial problems. Though officially independent, all three are funded and supported by POSCO, so the bond is strong enough to yield desirable results.

Summary Profile

In the public sector, the main interest is in the efficient utilization of limited government R&D expenditures, which total is one-third of the expenditure by private industry. Since this expenditure is from tax payer–supported public funds, the government feels the pressure. Another aspect is that the public sector wants to be the first to take a certain risk or to face the challenge of sailing to the unknown world. To be in the vanguard, to search for a new way to solve a problem, is part of the mission given to the public institutions. In the academic sector there is more interest in theoretical research, yet practical research is also important. In this sense MOT is a good way for an academician to work with an industrialist. Many academicians work as consultants to industry, stressing the importance of MOT to the managers. In Confucian culture, scholars enjoy a certain social status and respect. Once they are known as experts in MOT, their influence may go far beyond their real capacity to instruct either the public sector or the private sector. In the private sector, the biggest concern is whether an investment produces a return; if MOT makes a positive contribution in this regard, the interest in it can be tremendous. Since Samsung is known for its managerial skills and know-how, its interest in MOT has an impact on industries. Thus the influence of leading companies and the government may be an effective way to demonstrate the need for the application of MOT skills to many institutes and fields.

The Prevailing Influence on MOT

There seem to be two approaches for MOT, namely the Japanese approach and the American one. The Japanese approach is more related to application, to market-oriented technology and product development, while the American approach is more related to science-oriented and research-oriented MOT in general. In MOT within industries, the Japanese approach is more often emphasized, while government-supported research institutes are more accustomed to the American approach.

Historically China was Korea's main suzerain state. This changed early in the twentieth century. Japan colonized Korea in 1910, and then Japan was defeated in World War II. Since the war, the United States has been the most influential country in Korea, though Japan is also a powerful economic influence. But in MOT, particularly in industry, the economic status achieved by the Japanese in the past thirty years has had a tremendous impact on Korean industries. Thus in the private sector at least, the major influence on MOT is believed to come from the Japanese experience, though this does not necessarily exclude the influence of other countries. Although the university faculties are predominantly American-trained and there have been numerous foreign-aid programs from such international agencies as the World Bank and the Asian Development Bank, the need for MOT comes from an internal awareness created by economic and technical competition in international markets. Although MOT has not yet become the most important aspect of Korea's economic development, there is a growing awareness of the need for technology management skills.

Policy Implications

The fact that MOT has become an important managerial issue is by itself desirable for the development of Korea. Government, business, and the population at large have long been interested in hardware production; it was not until recently that there has been a recognition of the value of software, whether it be computer software or software in the sense of managerial expertise. Intellectual property rights, pollution, trade pressure to open the Korean market, and other pressures have forced policy makers, businesspeople, and academicians to change their way of thinking. The competition is fierce. This is true not only of international trade but in all business and managerial fields. Education is also highly competitive. Those without competitive minds and abilities will be doomed to failure. In academic society, the pressure to "publish or perish" seems comparable to heated battles in the marketplace. Competition affects everything: R&D for new products that are economically viable, teaching and research in school, businesses seeking the top managers, and policy makers and implementers trying to adopt the right industrial policy for economic development. MOT is an important aspect of technological development, vital for survival in this competitive world.

7

Management of Technology in Thailand

Chatri Sripaipan

Background

The rapid pace of technological development, the globalization process, and environmental issues have made people more aware of the role of technology as a factor determining international competitiveness and sustainable development. Technology not only needs to be generated, acquired, utilized, adopted, and adapted; like a business, it also needs to be managed. In a developing country like Thailand, the pattern of MOT is rather confused. It is not clear whether anyone is trying to manage technology. This chapter, then, is a first attempt to describe the emerging pattern. It is divided into three parts: The first part provides background on the three main productive sectors, the development through seven five-year plans, and the history of S&T development and policy. The second part describes the strategies of the three main players in technology management: the public sector, academia, and the private sector. Third, some policy implications are derived.

Profile of the Thai Economy

Agriculture

Thailand has an area of 513,115 square kilometers and as of 1992, a population of 57.6 million people. Lying in the heart of Southeast Asia and with five major rivers irrigating its vast central plain, Thailand has been a land of abundance and an important rice-exporting country for centuries. In fact, the old Thai saying, "there is fish in the water and rice in the field," remained true until recently. Today, 62 percent of the workforce is employed in the agricultural sector. Although the growth of employment in the industrial and service sectors has been much higher, the agricultural sector will probably retain its importance to the Thai economy for the foreseeable future in terms of employment and input into other sectors, as 25 percent of manufacturing industry is related to agriculture. Despite the fact that the share of GDP attributable to the agricultural sector has declined from 23.2 percent in 1980 to 11.8 percent in 1991, and that its growth rate is lower than that of other sectors, the value of agriculture is still growing in real terms. To maintain Thailand's international competitiveness, agriculture-based exports have been diversified, with such high value-added products as cultured tiger prawn, poultry, and fruit and vegetables being added to more traditional exports like rice, rubber, and tapioca (BOI 1992a, 1992b).

Manufacturing

Thailand's rapid transition from an agriculture-based, land-surplus economy into an industrializing country began only decades ago. Before 1930, there were only rice mills, sawmills, and cottage industries. The first industrial development policy was formulated after the change from absolute monarchy to constitutional monarchy in 1932. Interestingly, that policy had strong nationalist overtones; it restricted foreign ownership of industry and created a large number of state-owned enterprises with the purpose of generating revenues and encouraging Thai entrepreneurship. This policy continued into the 1950s, when the government was operating such industries as rice milling, sugar refining, textiles, ore smelting, rubber, cement, paper, and tobacco. More often than not, these operations were not profitable because of poor management. As a result of the change of government in 1958 and partly because of a strong recommendation from the World Bank, the government shifted away from establishing industrial ventures to promoting private-sector investments (Sibunruang 1986).

Another milestone in the industrial development policy is the establishment in 1960 of the Board of Investment (BOI), chaired by the prime minister. BOI seeks to promote industrial investment by offering incentives such as tax holidays, exemption of import duty on machinery and raw materials, and permission to own land and to bring in foreign experts and entrepreneurs. In the 1960s, the emphasis was on investments that utilized domestic raw materials and produced consumer durables for import substitution. In the 1970s, the policy shifted toward export orientation, but it did not really succeed until the 1980s. The Plaza Accord of 1985 has brought in a large influx of foreign investments since 1987. This flow has tapered off somewhat in the 1990s, but it remains high when compared to the pre-1987 years.

At present, Thailand is exporting a wide variety of manufactured goods, ranging from canned seafoods, gems and jewelry, textiles and garments, leather products and sport shoes to electrical and electronic consumer goods, computer peripherals and integrated circuits, and automobiles and auto parts. Heavy industry has also been developed. The eastern seaboard, about two hundred kilometers east of Bangkok, houses a huge petrochemical complex producing a great variety of petrochemical products made from natural gas from the Gulf of Thailand. A hot-rolled and cold-rolled steel plant is being constructed, with plans for it to eventually become fully integrated.

The industrial sector recorded the highest growth rate throughout the last three decades. The sectoral share of GDP attributable to manufacturing has increased from 21.3 percent in 1980 to 26.6 percent in 1991.

A study conducted by the Thailand Development Research Institute (TDRI) in 1988 concluded that of the four technological capabilities[1] in Thai industry, operative capability is generally the highest, followed closely by acquisitive and adaptive capabilities, while there tends to be a very low level of innovative capability (TDRI 1992a). In Wong Poh-Kam's conceptual framework (1992) on technological capabilities at the firm level, Thailand could be characterized as having low product technology, medium process technology, and low market/services.

Trade and Services

Trade and services include transportation and communications, trade, banking, insurance, real estate, public administration, and other services. The remaining sectors of the GDP (excluding agriculture and manufacturing) are mining and quarrying, construction, and utilities. The trade and services sector has the largest share of GDP, 45.8 percent in 1980 and 48.8 percent in 1991. Its rather constant share in the economy suggests that the trade and services sector has continued to be an important factor for the growth of the overall domestic economy despite the fact that the tourist industry, with 5.1 million visitors in 1991, has been the number one foreign exchange earner for some years.

The National Economic and Social Development Plans

In 1961 Thailand formally entered the "era of development" by announcing its First National Economic and Social Development Plan. Since then, six five-year plans have elapsed, and currently the seventh plan (1992–96) is in effect. Each plan has evolved and changed over time to address current developmental issues. However, these plans do not look beyond a time horizon of five years and therefore lack a long-term visionary approach to national development. On the other hand, the responsibility for translating these plans into action programs rests in the hands of relevant government ministries. Many times the programs have been delayed or diverted due to bureaucratic inertia, conflicts among agencies, or new political directives. The lack of an effective monitoring mechanism is another factor hampering both planning and implementation. Nevertheless, it is only fair to say that these plans have served their purpose and have played an instrumental role in guiding the direction of national development (NESDB 1992; Tanghongs 1992).

The First Development Plan (1961–66)

The first plan furthered economic development mainly by undertaking major infrastructure projects such as highway construction, construction of dams for irrigation and hydroelectric power, and construction of thermal-electric power plants. Due to favorable conditions both domestically and internationally, the economy achieved an average growth rate of 8 percent during the period.

The Second Development Plan (1967–71)

The development policies of the first plan continued under the second, but it also proposed the wider distribution of development benefits, particularly to the poorer rural areas. The economy still grew at a high rate of 7.5 percent per annum. Apart from development programs of concerned ministries, a number of special projects for farmers, such as the Provincial Development Project, the Accelerated Rural Development Project, and the Assistance Project, were launched. More attention was devoted to social development, particularly education and public health.

The Third Development Plan (1972–76)

Under the third plan, disparities of income and development that resulted partly from the success of the first two plans began to be addressed. To reduce differences in income levels of and services to the population, the plan sought to reduce the population growth rate through family planning and to create fairer distribution of economic and social services by developing agricultural and credit institutions and by stabilizing the prices of primary products. However, this was a period of turbulence both internationally and at home. The U.S. dollar was devalued, and skyrocketing oil prices followed. At home, the military government was overthrown by popular revolt, which was followed by a succession of weak democratic governments. Nevertheless, the economic situation improved toward the end of the plan, and Thailand attained an average growth rate of 7.1 percent per annum during this time.

The Fourth Development Plan (1977–81)

This was the first time that the National Economic and Social Development Plan was changed from an allocative plan to an indicative plan, which provided ministries with more flexibility. The fourth plan focused on two aspects of development: economic recovery and the development of natural resources. Though the economy was able to grow at an average rate of 7.1 percent per annum, a recession in the world economy and the rising cost of imported oil made the trade deficit rise to 7.6 percent of GDP. Government spending in the first two years of the plan was too high, causing a price-level increase of 11.7 percent per year. Moreover, income and social services were not broadly distributed. Though the proportion of rural poor in the total population had been reduced from 39 percent to 34 percent in 1976, it was still very large, representing more than ten million people.

The Fifth Development Plan (1982–86)

This was a period of low economic growth. GDP grew at only 4.9 percent per annum, though this was still higher than the world average of 2.5 percent. International debt, the trade deficit, the negative balance of payment, and the inflation rate were at record highs. The government had to impose monetary and fiscal restrictions by implementing a number of measures, such as a strict evaluation procedure on large capital expenditures, loan ceilings for government agencies and state enterprises, and limitation of commercial bank loans. To stimulate the economy and exports, measures such as tariff reduction and currency devaluation were introduced. Though these prescriptions caused hardships for some industries and businesses, they did work to stabilize the national economy, enabling Thailand to take advantage of the booming world economy during the sixth plan.

The Sixth Development Plan (1987–91)

This was the period of highest economic growth, with GDP growing at an average rate of 10.5 percent a year, and twice exceeding the plan's target rate. As a result of the fiscal and monetary restrictions imposed in the fifth plan, all the macroeconomic signs had improved—a revenue surplus, a surplus balance of payment, and a much-reduced foreign debt were achieved. With high growth rates in exports, investment, and tourism, the economy had become more global, with international trade increasing to 80 percent of GDP in 1991. However, there were still a number of imbalances in the overall development pattern. Income disparity continued on an upward trend. The rapidly growing economy had caused severe bottlenecks in many basic infrastructure services. The deterioration of natural resources and the environment was becoming evident. There was a severe shortage of skilled technical personnel, and the rapid development of the private sector seemed to leave the public sector behind.

The Seventh Development Plan (1992–96)

The seventh plan was designed to address the earlier imbalances. Though one of its three major objectives was to "continuously sustain the country's economic growth at an appropriate level, with stability," the other two had clear social goals: "to distribute income and development prosperity to the regions" and "to develop human resources, quality of life, environment, and natural resources."

Despite the Gulf War, the world recession, the coup d'état in February 1991, and the tragic political confrontation in May 1992, the Thai economy still grew at 7.9 percent in 1991 and 6.9 percent in 1992. It is expected that the growth rate will continue to increase at a rate not below 8 percent until the end of the plan (TDRI 1992b). The volume of investment and exports is expected to grow at a high rate. Infrastructural bottlenecks should ease toward the end of the plan due to investment in large infrastructure projects. It is also hoped that the various government efforts in developing regional areas will show some results.

History of S&T Development and Policy

Throughout history, the Thai people have been able to acquire many advanced technologies of the day, adapting them for use under local conditions without making a clear distinction between imported and invented technologies. King Ramkamhaeng of the Sukothai era more than seven hundred years ago imported celadon technology from China. Since the middle of the Ayuthya period about three hundred years ago, many modern European technological advancements, such as firearms and plumbing systems, have been introduced. During the Ratanakosin (Bangkok) period (A.D. 1782 to the present), Thailand has always been one of the first countries in this region, if not the first, to import the latest invention: steamships, railways, automobiles, telegraphs, telephones, radio, and television, among others.

Thais claim that their own inventions include the tricycle, walking tiller, long-tail boat, and "tut-tut" motortricycle. However, the inventors were never concerned about proper documentation, which makes it impossible to prove their origin. A more recent example is word-processing software for the Thai language. The two most popular programs are CU-Writer of Chulalongkorn University and Word Rajavithi of Rajavithi Hospital. Each group is still developing newer versions of the software, even though both programs are being distributed free without copyrights.

The above reference to historical cases is to illustrate the general attitude of the Thai people toward technology and invention, not to justify the current violations of international property rights (IPR) by some Thai

with scholarships for higher education. Lastly, it was the first time that the target for R&D expenditures was set at 0.5 percent of GDP, though actual expenditures as a percentage of GDP were 0.29 percent in 1982 and fell to 0.13 percent in 1986 (TDRI 1990).

The sixth plan (1987–91) was devised to muster the strength of government agencies and build up S&T capability in order to absorb new technologies and to enhance technological self-reliance. Highlights of this plan included the activities of STDB and the three technology centers at MOSTE, which led to sharply increased R&D in public universities. In addition to the Revolving Fund for the Research and Development of Technology, two more public sector loan funds were established by STDB and the Bank of Thailand. The Board of Investment also started to promote R&D ventures, providing quite favorable incentives as compared to ordinary industrial investments. On the human resource side, severe shortages in the S&T workforce prompted the Ministry of University Affairs to increase the enrollment of engineers in fields where the need was greatest by about 50 percent in 1988, to establish new engineering programs in Mahidol University and Thammasat University, and to create the new Suranaree University of Technology in northeastern Thailand. In addition, MOSTE had a program to send 1,200 students from the undergraduate to the doctoral level to study abroad in S&T fields to boost R&D human resources in research institutes and universities.

The seventh plan shifted the emphasis of S&T development from the public sector to the private sector (Tanghongs 1992; NESDB 1991). The hope is that ultimately Thailand's productive sectors will have the technological capability to compete internationally. The programs of the public sector have to be oriented toward linking with and upgrading the private sector. The plan set specific targets to substantially increase human resources in S&T, as shown in Table 7.1. The target R&D expenditure was increased to 0.75 percent of GDP, including 0.5 percent from the government and 0.25 percent from the private sector. Another feature worth mentioning is that the plan targeted industries at sectoral levels for the first time. These industries, crucial for the development of Thailand in the next five years, are electronics, metalworking and machinery, petrochemicals and plastics, textiles, food, gems and jewelry, and iron and steel. So far, most significant is the establishment of NSTDA and the Thailand Research Fund (TRF), which, although funded by the national government, operate as autono-

Table 7.1

S&T Personnel Targets: Seventh Plan

Category	1990 Per 10,000 People	1990 Total	1996 Per 10,000 People	1996 Total
S&T researchers	1.4	7,852	2.5	15,251
Scientists	7.2	40,379	10.2	62,225
Agriculturalists	6.7	37,575	10.5	64,055
Engineers	9.8	54,960	14.9	90,897
Technicians	141.5	793,560	221.5	1,351,260

Source: NESDB (1991).

mous juristic entities outside the civil service, and are able to attract the needed high-quality human resources and flexible enough to perform their duties with high efficiency. In particular, the TRF was designed to provide the main source of R&D funding for the country by accumulating an endowment fund on the order of 10 billion baht and dispersing funds to researchers in all fields in both the public and private sectors.

The MOT Challenge

As Thailand moves toward a more open economy, all its productive sectors will face severe competition in the world market. The agricultural sector can no longer rely on the "home-grown technology" of traditional crops; neither can the industrial sector rely on ready-made imported technology, nor the service sector on barriers that restrict foreign companies. The challenge is, therefore, to develop a new system of technology management drawing upon the complementary capabilities of the government, academia, and the private sector. At the national level, the four main functions of MOT as suggested by the United Nations are essential: strategic planning, organizing, monitoring, and leading; essential as well is an information system adequate to support decision making (Minden and Wong 1992). Though the seventh plan is clear in its directives to increase the effective use of technology in the private sector, the challenge is to implement the functions of organizing, monitoring, and leading, as well as to develop sufficient high-quality human resources. At the firm level, entrepreneurs and managers are to become more technology conscious in a more competitive environment. The challenge here is to provide the firms, especially small and medium-sized ones, with access to technology at a reasonable cost. To achieve this, the infrastructure needed for information services, consultancy services, technical services, training, technology licensing, and contract R&D, as well as supporting industries, must be built up to allow the firms to make more independent decisions regarding technology acquisition. In terms of TDRI's conceptual framework on technological capabilities, firms are to strive for better acquisitive and adaptive technological capabilities, which in turn will improve their operative capability. Therefore the MOT skills required will be technology assessment and negotiation (for price, terms, and conditions) as well as technology digestion and minor product and process modification.

Technology managers must be able to relate S&T concerns to the economy or business—in other words, to bring S&T into the broader context of planning, assessment, and management. This skill cannot be acquired through separate education in the technical field and the business or social science fields. Rather, MOT courses have to integrate the concerns of both sides. On-the-job training is another essential aspect of MOT skill development.

Organization of Management of Technology

There is no formal organization of MOT in Thailand. Strategies are uncoordinated both within each sector and among the three.

Public Sector Strategies

Public sector strategies are implemented on two levels: the national level, which is the responsibility of the National Economic and Social Development Board (NESDB), and the agency level, with various ministerial agencies responsible for carrying out specific tasks. The NESDB is entrusted with the task of formulating five-year economic and social development plans, coordinating interministerial policies, and evaluating projects at the request of the government. Its Technology and Environment Planning Division is assigned the following functions:

- to analyze and propose science, technology, and environmental policies;
- to prepare the Science and Technology Plan and the Environment Plan for the National Economic and Social Development Plan;
- to coordinate the preparation of action plans for S&T and for the environment on both a one-year and five-year basis, in line with the National Development Plan;
- to analyze science, technology, and environmental projects and plans;
- to develop databases and monitor the progress of the projects and plans;
- to participate in or support the operation of concerned agencies.

In addition, the division would like to assess the status of technology in Thailand. It identifies the following functions of public sector MOT:

Technology Transfer

Public sector MOT should facilitate technology transfer through foreign investments (whether foreign direct investments or joint ventures) in Thailand. The office of the Board of Investment (BOI) implements this policy by encouraging foreign companies in Thailand to promote Thai personnel to higher positions, to refrain from importing machinery already manufactured domestically, and to find a local source for parts and components. Recently, BOI has set up the BOI Unit for Industrial Linkage Development (BUILD) to help principals and suppliers in Thailand link up.

Technical Assistance

Public sector MOT should provide technical assistance to Thai industries through the Department of Industrial Promotion (DIP) of the Ministry of Industry (MOI), and the Department of Science Service of the Ministry of Science, Technology and Environment (MOSTE). Additionally, a number of training courses are provided by several divisions of DIP, such as the Metal and Machinery Industries Development Institute (MIDI), the Thailand Management and Productivity Development Centre (TMPDC), the Textile Division, and the Industrial Service Division.

Human Resources Development

Public sector MOT should provide technical personnel through formal education in universities and colleges as well as through a number of short-term training courses for working people. The Department of Skill Development in the Ministry of Labour and Social Welfare has nine institutes in various parts of the country to help graduates develop skills required for employment.

Research and Development

Public sector MOT should support R&D in government departments (e.g., the Department of Agriculture, the Department of Fisheries, the Department of Medical Sciences, the Department of Mineral Resources, and TISTR) and should provide funds through the National Research Council, NSTDA, and government funds budgeted to universities. The private sector can also access R&D funds from a number of sources mentioned earlier.

This view is very similar to a proposal by the World Bank, which suggested four strategies for MOT:

- the acquisition of foreign technology needed to reduce the gap between local best practices and international practices;
- the effective use and diffusion of technology, especially to reduce the disparity in economic efficiency among firms in different sectors and within sectors;
- the improvement and development of technology needed to keep up with the newest innovations;
- the development of a technical human capital base adequate to undertake the above tasks effectively (World Bank 1990).

A new public sector institution like NSTDA is interesting because it was created to make "a new and broad-based approach to science and technology development . . . by responding not only to technological advances but also to the changing economic and social context, . . . by interacting with other components of the national S&T system, and by enhancing and accelerating technical change and capability development in the production sectors" (NSTDA 1992). NSTDA has the opportunity to do the following:

- provide support for RD&E projects from research to commercialization or utilization by the public;
- provide support for technological upgrading and for development of the private sector;
- build up human resources and capability for RD&E in the strategic areas of science and technology;
- provide a site and a system for tripartite cooperation between the private sector (technology user), the universities (technology generator), and NSTDA (matchmaker and facilitator);
- use financial investment as a tool to develop science and technology (NSTDA 1992, 1993).

In its first year of operation, NSTDA provided RD&E support to universities and government agencies for a total of 64 projects in biotechnology, 83 projects in materials, and 32 projects in electronics and computers. It established and supported 14 laboratories and service centers, supported the purchase of 57 pieces of equipment for standards, testing, and quality control, and provided a large number of on-line S&T information services to universities, government agencies, and the private sector. Under the plan of Science and Technology for Commercialization, NSTDA provided three grants and loans to private companies, provided funding for fifteen industrial consultancy and technology development services, and formulated the master plan for the Science and Technology Research and Development Park. In human resources development, funding for 172 scholarships for overseas study, 36 scholarships for domestic graduate study, 23 technical training positions abroad, and 80 specialized technical training positions in Thailand was provided during NSTDA's first year.

Academic Strategies and Programs

The rapid economic growth and the fast pace of modernization in the private sector have left the academic sector on the defensive, since it is predominantly in the public domain. Higher education in the private sector has a rather short history and faces some of the same constraints its public counterpart does, such as a shortage of full-time teachers. The private companies, though, feel that the solution is for technology managers to employ graduates with M.B.A.s and with S&T degrees at the bachelor's level. Therefore, although Thailand has a large number of M.B.A. programs, the School of Management of the Asian Institute of Technology (AIT) offers an M.B.A. program in MOT (AIT 1992).

The twenty-month M.B.A. program in MOT is intended to train students in management with an emphasis on technological processes, projects, and resources. There are courses in subjects that fall into four broad groups: management tools and techniques, behavioral management and environmental analysis, functional management systems, and management of technological processes and resources. At the end of the courses, there is a thesis or research study option; topics are selected mainly from the field of technological systems management. Since the school started to enroll students in 1989, the classes have increased from about thirty students to about forty, some 60 percent of whom have an engineering background, with the rest graduated in sciences and social sciences. The students are about thirty years old, have five to seven years' working experience in both the public and private sectors, and represent about fifteen countries. Interestingly, students from Taiwan receive financial support from their companies and go back to the same companies after graduation. Students from developing countries who enter the program, including those from Thailand, typically have to resign from private companies, are supported by scholarship, and look for better jobs after graduation, although those from the public sector usually return to the same job. In one case a Thai student with a social science background resigned from the job of administrative executive of a manufacturing company to enter the M.B.A. program. Upon graduation, AIT customarily organizes student presentations and invited the student's former employer. This student's presentation on the "just-in-time" management technique impressed her former boss so much that he appointed her plant manager of the company.

Part of NSTDA's program for Professional Training Centres and Activities is the Technology Management Training Program, which NSTDA intends to implement in cooperation with the School of Management of AIT. Thus, the high-quality resources at AIT will be made more widely available to the Thai people.

The latest development in academic MOT is the establishment of a

Science and Technology Administration (STA) program at Mahidol University, which was introduced in July 1993. When fully developed, it will have three components:

- an executive development program: seminar series and short courses;
- undergraduate courses: elective courses for undergraduate studies in science and technology;
- graduate program offering a graduate diploma and a master's degree in science and technology administration.

Private Sector Strategies

To assess MOT in the private sector, three case studies are presented in three of the targeted industries of the Seventh National Economic and Social Development Plan.

National Petrochemical Corporation

The National Petrochemical Corporation (NPC) was established in February 1984 as a public/private venture to operate the first upstream petrochemical unit, or olefins plant, in Thailand. NPC buys ethane and propane gas from the gas separation plant of the Petroleum Authority of Thailand, which holds a 49 percent share. Ethane is cracked into ethylene, and propane is catalytically dehydrogenated into propylenes. The ethylene and propylene gases, called olefins, are then sold to four downstream units of the private sector which altogether own 45.8 percent of NPC and which produce plastic resins and resins such as HDPE, LLDPE, PP, LDPE, VLM, and PVC.

From the start, one of NPC's objectives has been to "bring natural resources to the people at a fair price." An executive team was set up to select the technology for the company that was in line with the business, had already been proven to a certain extent, was environmentally friendly, and was suitable for the available materials, the people, and the market. The top management chose the oleflex reactor system, the dehydrogenation system of Universal Oil Products, which had been proven for heavy hydrocarbons but never used for light hydrocarbons for propylene production. By taking a chance, NPC became the world's first commercial facility for this new propylene production technology. The decision was to some extent dictated by the availability of local raw materials. The technology developer guaranteed a minimum efficiency of 50 percent, but the plant can now operate at 80 percent efficiency.

The technology acquisition process emphasized the development of human resources, and active learning. Staff training stressed safety, reliability, quality, and efficiency. After only three years of operation, NPC was asked by ICI Corporation to train South Korean engineers for the oleflex reactor system in preparation for the start-up of their own plant. A technology transfer contract has been signed with a Thai company. Selling technology, of course, is not NPC's main business. It has been continuously trying to improve conversion efficiency by experimenting with new catalysts or new operating procedures emphasizing a high level of safety and environmental protection. NPC would like to see the government establish a national laboratory for calibration standards of gases.

Teijin Polyester (Thailand) Ltd.

Teijin Polyester (Thailand) Ltd. (TPL) is a joint-venture company pairing a major world polyester producer, Teijin Limited of Japan, with a group of Thai investors. TPL was set up in 1970 as a producer of staple fibers and filament yarns. During its first ten years of operation, the company concentrated on quantitative growth. The output of staple fibers and filament yarns expanded in that period from 300 and 120 tonnes per month to 2,500 and 1,200 tonnes per month, respectively. Due to the oil crisis and the erosion of Thailand's labor-cost advantage as compared to that of other developing countries, TPL has since shifted its focus to value-added products. However, TPL's emphasis has always been on the development of human resources. All TPL's employees receive on-the-job training at different levels. Managers, supervisors, and workers have the opportunity to learn new technology in Japan. At the same time, Japanese technicians are sent to transfer their experience to their Thai counterparts. Workers' initiative and feedback are encouraged through quality control circle activities and a suggestion system. As a result, the company claims to have significantly higher productivity than its local rivals and can remain competitive despite the fact that its workers receive a higher average salary.

In its early years, TPL received massive technical assistance from

Teijin of Japan. But within a few years it was able to achieve greater self-reliance with regard to plant operation and minor process modifications as well as significantly increasing its use of locally made machinery and spare parts. This is because in addition to its emphasis on developing the skills of its employees, TPL also works at upgrading local support industries, such as metal coating and plating and sandblasting. Moreover, it has created a working group composed of staff from various sections of the company, including production, maintenance, and instrumentation, which attempts to solve technical problems beyond the responsibility of the engineering division. The working group spent three years developing a new manufacturing process that uses terephthalic acid instead of dimethyl terephthalate, which was expensive and in short supply.

In the aftermath of the oil crisis, TPL was forced to restructure in order to raise productivity and move toward the development of value-added fibers. When the Technology Division was formalized in 1982, it was nicknamed "Cemetery for the Generals" because it was largely composed of surplus staff from the restructuring. However, with strong support from top management and under the leadership of the plant manager, the group was reorganized and reoriented to examine new product possibilities and local market potential. With virtually no laboratory facility, it started to work on the development of high value-added products. It had to request time, materials, and staff from the production division for its research. Its 1984 invention of the cationic dyeable process, which produced different shades in a single dyeing, was a milestone for the Technology Division, boosting its confidence in undertaking further product and process development.

When TPL decided in 1988 to invest in a new plant for its continuous polymerization process, the contribution of the Technology Division was paramount. In the new plant, 50 percent of the machinery was locally made, 90 percent of the construction was done by Thais, and 90 percent of the start-up work was done by local staff. Had TPL not paid attention to developing its indigenous technological capability, much of this work would probably have been executed by Japanese engineers.

Today, the Technology Division remains small and flexible, with the number of staff varying according to projects. All staff are borrowed from various divisions and are rotated according to project duration or every three years. The division's resilience depends chiefly on the support of top management and the understanding of fellow workers.

Technology Development Office, Premier Global Corporation

The Premier Group of Companies (PGC) is a conglomerate of about thirty companies in food, construction materials, electronics, and leasing. In 1990, the Premier Global Corporation was formed to represent the Premier Group. In June 1991, the Premier Global Corporation established a Technology Development Office (TDO) for the purpose of building up the R&D capability in PGC to meet the challenge of international competition. TDO is headed by a physicist who spent almost thirty years in the United States in the research laboratories of the General Electric Company.

Instead of investing in a research laboratory and equipment, as conventional wisdom would dictate, the efforts of the Research, Development and Engineering Center of TDO are focused on people and networking. The Food/Biotechnology Department inherited some food-testing activities from the food industries of PGC. Other departments had to find their own work. In the initial phase, a small group of capable technical personnel was employed to visit factories within PGC to identify problems that could be solved quickly, often by fixing or modifying existing equipment. The research team from the Equipment Development Department, therefore, has few in-house facilities but must quickly build job-specific tools and corrective parts, install them, and make them work. There has been a rapid increase in the demand for such a service. An electronic control to regulate the pressure of food steamers and a more uniformly heated oven that reduced the production loss of bathtubs, sometimes as high as 10 percent, to 1 percent are examples of TDO's successful technological innovations. Some of the forty TDO-supported senior-year projects in various universities have in fact been used to build parts for factories. The key reason these projects are ultimately applicable in industry is that they are designed and managed from the industrial side.

TDO has initiated many cooperative research programs with universities and research institutes. Premier's External Laboratory Program has established the Premier Applied Materials Laboratory (PAML) at Chulalongkorn University, is about to inaugurate the Premier Applied Physics Laboratory at Chiang Mai University, and is currently negotiating with several other universities. PAML, apart from receiving an annual 2-million-baht equipment grant, is also supported by TDO through a number of research projects. The External Cooperative Program has already established collaborative research projects with the Thailand Institute for Scientific and Technological Research (TISTR). These cooperative research programs serve to reduce "brain drain" from the public to the private sector by providing university staff with research projects and students with industrially relevant senior projects and thesis work.

Despite a down-to-earth approach in the initial stage, TDO has not forgotten that survival and prosperity in a competitive world require proprietary technology, so its Physics and Engineering Department attends to this area with in-house activities such as its ongoing research on electrophoresis. A memorandum of understanding between TDO and Chulalongkorn University calls for long-term proprietary projects to be carried out in the university with the Premier Group's researchers working alongside university staff. Royalties from patented technologies are to be shared.

While PGC continues to import most of the technology it needs, TDO has modified its existing technology and invented new technology, thus saving on costs and enhancing competitiveness. The Project Consulting Center of TDO is designed to form joint ventures and strategic alliances to induce technology transfer. TDO also has a Premier Students Program that awards scholarships to bright students in remote areas so they can complete their secondary education, and it sponsors the Premier Lecture Program for the Congress on Science and Technology of Thailand, publishes a quarterly technical journal, and supports various programs among the scientific communities.

Assessment

It is observed that a number of leading companies in Thailand are aware of the importance of technology to their operations. But none of them has yet developed a comprehensive MOT strategy. The three case studies presented above describe three firms' technological capabilities, which can be broken down into acquisitive, operative, adaptive, and innovative capabilities (TDRI 1992a).

National Petrochemical Corporation (NPC) quickly developed good acquisitive capability in order to ensure a proper start. Its MOT strategy was concentrated on technology assessment. Its heavy emphasis on the development of human resources resulted in its high operative capability. Its purchased technology has been so well absorbed that

NPC's staff can transfer it to other companies. To increase the conversion efficiency to 80 percent, NPC must possess a certain degree of adaptive capability. It is probably unreasonable to expect a three-year-old firm to embark on R&D of such highly complex and capital-intensive technology to develop its innovative capability. NPC has engaged a university professor to work on the catalysts and is waiting for the Petroleum Authority of Thailand to invest in a larger R&D facility.

Teijin Polyester (Thailand) Ltd. (TPL), being a joint venture, did not have to worry about its acquisitive capability at the outset, and in the early years its MOT strategy emphasized building up its operative capability. Its emphasis on human resources development made it possible to quickly develop its adaptive capability when needed. It is surprising that the Technology Division could develop new products with no laboratory facility of its own and virtually no permanent staff. Yet, it was able to identify its own niche and demonstrate its innovative capability. This success must be attributed to capable personnel management.

The Technology Development Office (TDO) of the Premier Global Corporation (PGC) probably represents a more aggressive MOT strategy for technology development. TDO is in a way similar to the Technology Division of TPL; both solve technology problems in production. But with its permanent staff, laboratory facility, and management flexibility, TDO has been able to work on longer-term and riskier projects. Not only does it have medium- and long-term projects, but it funds many cooperative research programs in universities and research institutes. TDO has been called the NSTDA of the private sector. However, even the director of TDO conceded that it will take some time before PGC's business strategy and its technology strategy can be integrated.

The Role of Multinational Companies

Multinational companies (MNCs) occupy a much more prominent role in the development of the industrial sector than in the agricultural and service sectors, due to the government's industrial investment promotion policy. Initially, MNCs came to Thailand in the 1960s to enjoy the protected domestic market. Nowadays the majority of these companies are export-oriented and operate in Thailand because of low-cost land and labor and low overhead. It is not surprising to find some new

MNCs that have world-class operative capability but whose other technological capabilities are negligible because they can rely on the main company for them. Over the years, most MNCs develop significant acquisitive and adaptive capabilities and in some cases even some innovative capability. TPL is an exemplary MNC that has developed good technological capabilities. Its dependence on the mother company's technology has been greatly reduced by finding sources of needed technology worldwide.

Mechanisms for Interaction

A number of linkages exist among the private, public, and academic sectors, though they are generally not directly concerned with MOT. This chapter has already discussed various governmental agencies concerned with technology transfer, technical assistance, the development of human resources, and R&D, presumably all for the benefit of the private sector. The private sector may view this differently. The Board of Investment is generally seen as the agency to give privileges and incentives to industrial investments. The development of an adequate workforce is seen as the duty of the government without any need for active participation from the private sector. Technical assistance and applications of R&D results to the productive sectors are exceptions rather than the general rule. The National Science and Technology Development Agency (NSTDA) has a number of programs to interact with industry and universities such as industrial consultancy and technology development, private-sector RD&E support, technology transfer and technology development financing, a Science and Technology Research and Development Park (STRDP), and scholarships and visiting fellowship programs.

Though the Asian Institute of Technology's M.B.A. program in MOT is popular, the number of openings for Thais is very limited. Thai institutions still find it easier to run a standard M.B.A. program than to offer MOT courses. It is reported that interaction between universities and industry in Southeast Asia in general and Thailand in particular is concentrated mainly on providing training courses and consultancy services to industry with a number of collaborative research programs and a few commercial applications of university research results (Sripaipan 1992). TDO of PGC is a private sector initiative that can bring about better interaction between universities and industry.

Policy Implications

For the Public Sector

There is need for public policy and planning at the national level. In fact, a National S&T Policy Agency has been proposed (TDRI 1991). If such a proposal is not accepted, the Technology and Environment Planning Division of the National Economic and Social Development Board (NESDB) can still assess the status of S&T, analyze and propose policies, and monitor the progress of projects and plans. Once MOT is understood and accepted at the national level, planning at sectoral levels is also possible.

The technical assistance provided by various ministries has yet to produce a significant impact on industrial development, especially in the small and medium-size industries that they are supposed to serve. Some quantitative targets will have to be set in proportion to available personnel, facilities, and funding.

A strong commitment will have to be made to provide a realistic solution to the shortage of skilled personnel (not only in S&T fields). A quantum jump in expenditures for education as a percentage of GDP is essential if the problem is to be solved in the long run (Sripaipan and Brimble 1992).

Funding for R&D should be demand-driven rather than supply-driven (TDRI 1991), which would link it more closely to users. Basic research should be funded separately.

NSTDA, being broad-based and flexible, can implement many of the above suggestions, but its work will have more impact if it is done in cooperation with other governmental agencies.

For Academia

Though the academic sector is not supposed to set policy, one important benefit in freeing the universities from the civil service is that they would have to be more responsive to the needs of society. They would have more liberty to establish new courses, since such decisions would be made by university councils instead of by the Ministry of University Affairs. It is hoped that the recent Science and Technology Administration program is successful and that more MOT courses will follow in Thailand.

For the Private Sector

Competition is one sure way to stir up the innovative instincts of entrepreneurs. They know that in a competitive environment they have to make proper use of all their resources, including technology, by efficient management. To make entrepreneurs aware of the threats and opportunities, a variety of trade associations can be helpful in bringing in up-to-date information and technology. Such "serious" associations are lacking in the Thai private sector.

Note

1. Technological capability at the firm level was broken down into four classifications: *acquisitive capability*, a firm's ability to search, assess, negotiate, and procure needed technologies, and to install and start up production facilities; *operative capability*, its ability with regard to the operation, control, and maintenance of production facilities, as well as to skill development, production planning, and quality control; *adaptive capability*, a firm's ability to digest technology and to make minor product and process modifications; and *innovative capability*, the capacity to make radical product and process modifications, to carry out in-house RD&E, and to invent new products and processes.

References

AIT (Asian Institute of Technology). 1992. School of Management brochure. Bangkok: Asian Institute of Technology.

BOI (Board of Investment). 1992a. *The Investment Environment in Thailand*. Bangkok: Office of the Board of Investment, Office of the Prime Minister.

BOI 1992b. *Key Investment Indicators of Thailand*. Bangkok: Office of the Board of Investment, Office of the Prime Minister.

Minden, K., and Wong Poh-Kam. 1992. "Developing Technology Managers: Comparative Pacific Rim Strategies." Concept paper. Conference on "Developing Technology Managers: Comparative Pacific Rim Strategies," 11–12 March, Centre for Management of Technology, National University of Singapore.

NESDB (National Economic and Social Development Board). 1991. *The Seventh National Economic and Social Development Plan: 1992–1996*. Bangkok: Office of the National Economic and Social Development Board, Office of the Prime Minister.

NESDB. 1992. *From the First to the Sixth Development Plan: Key Issues and Results*. Bangkok: Development Studies and Information Division, Office of the National Economic and Social Development Board.

NSTDA (National Science and Technology Development Agency). 1992. *Strategic Plan: 1992–1996*. Bangkok: National Science and Technology Development Agency.

NSTDA. 1993. *Annual Report: 1992*. Bangkok: National Science and Technology Development Agency.

Siburuang, Atchaka. 1986. *Industrial Development Policies in Thailand*. Report prepared for the World Bank.

Sripaipan, C. 1992. "Financing Industry-University Interaction: A South-East Asian Perspective." Paper presented at the Technology, Commercialization and R&D International Workshop on Industry-University Interaction (organized by the International Development Research Centre, Canada), 24–26 June, Bangalore, India.

Sripaipan, C., and P. Brimble. 1992. "Science and Technology Manpower in Thailand: Toward a Long-term Solution." In *Manpower Planning for the Development of Industrial and Services Sectors*, submitted to the National Economic and Social Development Board by the Thailand Development Research Institute.

Tanghongs, Thippaya. 1992. "Technology Development in the Seventh Development Plan." *Economical Social Journal* of the National Economic and Social Development Board 29(5): 35–42.

TDRI (Thailand Development Research Institute). 1990. *The Baseline Data Project: S&T Infrastructure and Development in Thailand*. Bangkok: Thailand Development Research Institute.

TDRI. 1991. *The Management of the R&D System to Support Industrial Development*. Bangkok: Thailand Development Research Institute.

TDRI. 1992a. *The Development of Thailand's Technological Capability in Industry*. Research monograph no. 9. Bangkok: Thailand Development Research Institute.

TDRI. 1992b. *Thailand's Macroeconomic Outlook: 1992–1996*. Bangkok: Macroeconomic Policy Program, Thailand Development Research Institute.

Vechamaneesri, Boonyong. 1990. "Directions of Development in Thailand." Private communication.

Wong P. K. 1992. "Development of National Technological Capabilities in Singapore: Past Strategy, Future Management Challenges." Paper presented at the S&T Task Force workshop on "Integrating Technology and Management: A Tripartite Approach," conference of the Pacific Economic Cooperation Council, 9–10 November, Jakarta.

World Bank. 1990. *Technology Strategy and Policy for Industrial Competitiveness: A Case Study in Thailand*. The World Bank Industry and Energy Department. Industry Series paper no. 24.

8

Developing Technology Managers: The Malaysian Experience

M. Rahimah Said

Background

In the early five-year development plans, initiated after Malaysia's independence, the emphasis was on rural development and agriculture. But as the economy matures and development of the nation begins to hinge upon the dictates of the world economy and the global marketplace, some choices have to be made. The imperative for Malaysia in the 1990s is to industrialize in order to become an advanced, affluent nation. The major challenge is to balance development between urban and rural areas, between industry and agricultural activities. The choices are complex and not mutually exclusive. But clearly, Malaysia's agricultural sector must be fully strengthened, for it is one of the basic pillars of the national economy; after all, 40 percent of Malaysia's GNP comes from this sector. Equally clear, Malaysia must industrialize.

The Industrial Master Plan (IMP) of 1985 reiterated the vital role industrial progress will play in the sustained economic growth of the country. It represents a watershed in the structural shift of the country's economy from one dependent largely on agriculture and primary products to one in which the manufacturing sector plays a much bigger role. It also signaled a change to a distinctively target-oriented approach to industrialization, from one largely laissez-faire. The focus became export orientation. In 1984, the manufacturing sector became the largest contributor to GDP for the first time. In 1990, it contributed about 25 percent of GDP and it accounted for an estimated 34 percent of new jobs in 1989. Thus manufacturing is a critical sector of the Malaysian economy and is growing in importance.

The IMP recognized and stressed that the active promotion of technology was needed so that a strong foundation for industrialization could be built. Technology provides the means to achieve the transformation of available inputs into desired outputs. It allows value enhancement through the process of transforming primary products into semifinished, consumer, and capital goods. Technology has demonstrated an extraordinary capacity to generate wealth and power. Industry without technology is like a vehicle without an engine. Today, the capacity to use and develop technology through innovation, the management of information, and the nourishment of knowledge as a resource will determine economic authority, as natural resources, capital, and labor once did.

The issues facing the Malaysian industrial sector highlighted the need to improve labor productivity, to expand high value-added outputs that are technologically oriented and strongly support-oriented, and to diversify export products and markets. Faced with rising labor costs and competition from other low-wage countries in the region in the traditional export of manufactured goods, Malaysian industries need to rapidly upgrade their technological sophistication and product quality in order to remain internationally competitive.

The low science and technology base in Malaysia is a big constraint to broadening industrial activity. Large-scale industries that are most able to undertake R&D activities and to transfer technology operate mostly in enclaves. As a result, innovative activities are geared largely toward incremental and marginal improvement rather than the design and development of new products. The problems of poor quality and high costs handicap industries making use of the products of small and medium industries (SMIs) as intermediate inputs. Parallel to the development and enhancement of the country's technological base, the government needs to give greater attention to the establishment of vertical and horizontal linkages within the various manufacturing subsectors. The SMIs are expected to provide the much-needed broadening and deepening of the industrial base. Expansion in the form of a diversified industrial structure consisting of a multitude of modern SMIs will provide support and technological linkages. This will encourage the larger manufacturing industries to locate their ancillary and subassembly supplies domestically. In addition, the technological development of SMIs is impeded by the low technological level of their managerial staffs and by shortages of skilled and scientific personnel. The R&D and related technology infrastructure needs to be developed to provide the necessary support for the technological development needs of SMIs.

National Science and Technology Policy

Historically, the development of science and technology in Malaysia reflects early developmental needs and priorities and their comparative advantages. About half the research establishments are in agriculture or related sectors. The earliest research institutes (RIs) and public funding for R&D in Malaysia date back to its preindependence period, when the Institute of Medical Research, the Rubber Research Institute, and the Forestry Research Institute were established. Subsequently, several RIs were established to stimulate R&D and to provide technical advice and service to various sectors of the economy. RIs in agriculture and related sectors employ the bulk of R&D personnel, followed by universities, reflecting the longer history of these establishments.

A conscious effort by the government to stimulate R&D in the industrial sector is relatively new in Malaysia compared to its long experience with the RIs in the agriculture sector. The establishment of the Ministry of Science, Technology and the Environment (MOSTE) in 1973 marked a significant step by the government toward effective coordination and planning of the development of industrial technology. MOSTE is at the apex of the science and technology (S&T) framework, and is assisted by the National Council for Scientific Research and Development (NCSRD).

Malaysia's science and technology framework is a decentralized system, with policy formulation and operational responsibilities shared among a number of agencies. MOSTE and the Economic Planning Unit of the Prime Minister's Department are the major policy-making agencies in the development of industrial technology, while sectoral ministries also play a role in their respective sectors. Research is undertaken within an overall planning framework as part of a five-year science and technology plan that complements the five-year national economic plan. The close links between the two plans ensure that the science and technology plan reflects the country's economic and investment requirements of S&T. MOSTE is responsible for the formulation and implementation of a national science and technology policy, and for identifying priority R&D areas, formulating recommendations on projects to be funded under the Intensification of Research in Priority Areas (IRPA) program, and prioritizing proposals for the country's five-year plans that cut across the responsibilities of individual ministries.

The status of industrial technology is intimately linked with the S&T policy framework as a whole. First, in the larger context, the priority assigned to S&T in relation to overall national goals and aspirations is of crucial importance, as it impinges both directly and indirectly on the development of industrial technology capability. The National Science and Technology Policy and the fifth and sixth Malaysia Plans emphasize the development of self-reliance in S&T in support of economic growth. Second, the relative importance accorded industrial technology within S&T policy has a direct bearing on the resources that can be assembled and deployed, and on the effectiveness of actions to develop national industrial capabilities. While there exists a structure of research institutes, public agencies, and institutions of higher learning with functions in various fields of science and technology, greater successes have been achieved in agriculture and the primary product areas, rather than the manufacturing sector and other secondary sectors of the economy. There is also little emphasis on applications-directed research. The level of industrial R&D expenditure, an indicator of the development of industrial technology, remains low and heavily skewed, with the public sector accounting for 90 percent. As in most developing countries, the development stage forms the weakest link in the chain, and this is where a body of expertise must be built up. Industrial development cannot take place in a technological vacuum. A major reorientation is called for: industry must recognize the need for R&D, and in turn, research bodies must emphasize practical industrial applications.

To an extent, existing policy instruments and institutional mechanisms have been less than adequate to foster industrial technology development, with little attention given to the generation of a minimum level of indigenous technology through product enhancements and innovation. There is a growing realization that market forces alone may not be sufficient, and that government must help to shape and promote technological development. Hence the S&T policy framework outlines the concept of shared responsibility for technological development, i.e., among the government, academia, and the business community. The challenge is to find the judicious mix, the golden mean.

The MOT Challenge

The following are of critical importance to Malaysia's industrial transformation:

1. The need to continue industrial technology development if Malaysia is to retain its competitive edge in export markets;
2. The need to progressively upgrade and restructure skill mix requirements to ensure that skills are relevant to the demands of an increasingly technology-intensive market;
3. The promotion of SMIs as suppliers of basic materials and parts to larger industrial enterprises and self-reliant export industries.

In May 1990, the government of Malaysia accepted the National Plan for Industrial Technology Development, known as the Action Plan, to further stimulate industrial R&D activities in the public and private sectors and to encourage linkages between R&D and industrial application. Initiated by MOSTE, this plan was conceived with active participation from academia and the private sector. The strategic thrusts of the Action Plan are the following:

1. Strengthening the institutional and support infrastructure for industrial technology development;
2. Ensuring the widespread diffusion and application of technology needed to enhance market-driven R&D and to adopt and improve new technologies;
3. Building competence for specialization in key emerging technologies;
4. Strengthening institutions and mechanisms for continual development and elevation of the technical proficiency of the human resource base;
5. Elevating S&T awareness and appreciation to provide the most conducive climate possible for invention, innovation, and technological advancement.

Organization of MOT

The Action Plan has forty-two comprehensive recommendations and when fully implemented will create a proper environment for continued technological development. Central to the implementation of this plan, however, was the considerable emphasis that was placed on human resource development (HRD).

Currently, the technological skills base in Malaysia is weak. This implies that the country is likely to face difficulties in absorbing and adjusting to newer technologies and manufacturing processes. It is essential, therefore, to focus on technical personnel planning in the industrial sector. The strengthening of the administrative mechanism for personnel planning and development became a priority for MOSTE.

Disciplines for technical personnel at the graduate and postgraduate level in Malaysia are determined at an early stage, during the selection of candidates. However, this has to be reviewed regularly to ensure that the needs for industrial technology are properly accommodated. Courses at institutions of higher learning need to be complemented by an emphasis on practical training, as proficiency in technical disciplines is acquired mostly through work-related experience. To provide effective support for the acquisition of competence in key new technologies, universities will have to build up their education and training capabilities. Teaching and research facilities in the universities must be expanded and upgraded to increase the number of science and engineering postgraduates in the country. With an expanded capability for teaching and research in new technologies, and with the existence of a broad base of multidisciplinary skills, universities should be well placed to contribute substantively to product development and the commercialization of new ideas. Centers with a critical mass of expertise in the new and emerging technologies need to be created to encourage linkages among universities, technology parks, and innovation centers, and measures must be instituted to allow mobility of expert personnel both to and from private industry and institutions of higher learning to enable them to participate fully in technological development.

To meet the objectives of the Action Plan, strategies for the HRD program were formulated through the following procedures:

1. Enrollment of the staff of departments, research institutions, and institutions of higher learning in specialized courses in industrial technology at foreign institutions and recognized centers of excellence;
2. Attachment of staff to foreign industrial organizations and S&T institutions;
3. Provision of in-country training for staff in various fields of science and technology to be conducted by foreign technology, specialists and training consultants;
4. Provision of university fellowships for postgraduate study at local and foreign universities.

Fields of study that provide hands-on experience have been designed to meet the needs of each institution's temporary attachments of researchers to foreign research institutions and industries. Participants will work in areas directly related to their fields to improve their level of expertise. Foreign technology and training experts also provide in-country training to enable considerably more participants to be included in the program.

The HRD program is centered on MOSTE, the Standards and Industrial Research Institute of Malaysia (SIRIM), the Malaysian Institute of Microelectronics Systems (MIMOS), the Malaysian Agricultural Research and Development Institute (MARDI), Palm Oil Research Institute (PORIM), the Nuclear Energy Unit, the National Institute of Public Administration (INTAN), the Ministry of International Trade and Industry (MITI), the Malaysian Industrial Development Authority (MIDA), the Economic Planning Unit (EPU), the Ministry of Finance, the Department of Environment, the University of Malaya, the Science University (USM), National University (UKM), Technology University (UTM), and the Agricultural University (UPM).

MOSTE, EPU, and the Ministry of Finance need skilled personnel to upgrade their policy-making capabilities and to collect, evaluate, and disseminate information related to R&D. The objective for HRD in research institutes is to teach and enhance skills through fellowships and training programs and to develop greater competence in industrial R&D. In the universities, as compared to the RIs, the need is for linkages with industries through exposure to new technologies. Sectoral ministries like MITI require in-house expertise in policy analysis in order to effectively participate in the formulation of national S&T plans and initiate actions for the industrial sector. INTAN, on the other hand, needed upgrading of its skills-based competence in industrial S&T, technology management, and R&D management to enable it to effectively provide policy and management training and courses in S&T. The following statistics pertain to the five-year program:

University Fellowships	560
Attachment to foreign institutions/organizations	94 (217 person months)
Short courses	801 (1,922 person months)
Experts	364 (432.75 person months)

Concomitant with the Sixth Malaysia Plan (1991–95), financial allocation under the development budget was committed for the HRD program. Though the figures represent only about 8 percent of the requirements, it is hoped this will provide the base for a stock of technological knowledge, expertise, and skills that will be built up in selected institutions and will gradually be disseminated to industry through technical services, product development, and personnel mobility.

At the private sector level, a Skills Development Fund (SDF) was launched in 1993, jointly managed by government and the private sector to finance industrial training and retraining programs. The Ministry of Human Resources and the Ministry of Education have been assigned the task of improving the certification system for technical personnel, expanding continuing education, enhancing linkages between industry and institutions of higher learning for the purposes of industrial training, and upgrading technical and vocational schools.

In addition to the SDF, which covers mainly medium and large industries, industry associations that are responsible for smaller businesses are assisted by MOSTE in meeting private sector requirements for training. MOSTE provides grants to enable training facilities to carry out approved programs. Initiatives for training by the private sector are further enhanced by joint programs between industry associations and state development organizations. A good example is the Penang Skills Development Centre. Equipment is donated by concerned industries, while facilities and other required infrastructure are provided by the state organization. Training modules are worked out together to meet local industrial needs. This center has been a model for other states to emulate. Programs have also been formulated to allow the private sector to use training facilities at technical schools and skills training centers in the country during the evenings. It must be stated, however, that this is a new program and that evaluation of its effectiveness would be premature at this stage.

Close cooperation and consultation is maintained between MOSTE, the lead agency for the management of technology, and the private sector, including industry associations. If, in the past, most research was carried out in isolation by RIs, a new approach has been adopted calling for consultation with industries on their needs, especially in R&D. More information is exchanged between the two groups, with the result that each has a better understanding of their mutual needs. The partnership of industry and research is no longer alien.

The role models for Malaysia have been Japan, for its benefits of consensus and participation, coordinated by MITI; South Korea, for the direct involvement of its head of government; Sweden, for its national consensus on social responsibility; and Singapore, for its top-down approach based on strong leadership with vision.

The HRD program must be complemented by the infrastructural changes needed to implement market-driven and application-oriented research. Organizational structures, compensation systems, recruitment processes, and other administrative procedures need adjustment. In other words, the organization must keep pace with the demands of modern technology and the needs of today's societies. Most government organizations are attempting to manage space-age technologies and serve modern societal needs with outdated methods. The mismatch has to be redressed. The issue that has to be addressed is change and the willingness to change. This means that workplace culture needs adjustment, too. In sum, if the organizational structure is not properly designed to cope with modern technological imperatives and environmental forces, then no amount of planning, organizing, directing, and controlling can compensate for it.

The prevailing organizational structure of government research institutes can be said to be "classical," based on a division of functions that enables them to respond to local diversity and variations. In this type of organization, division managers have the resources and power

Figure 8.1. **Special Project Structures Are Needed in a Contract Research Organization**

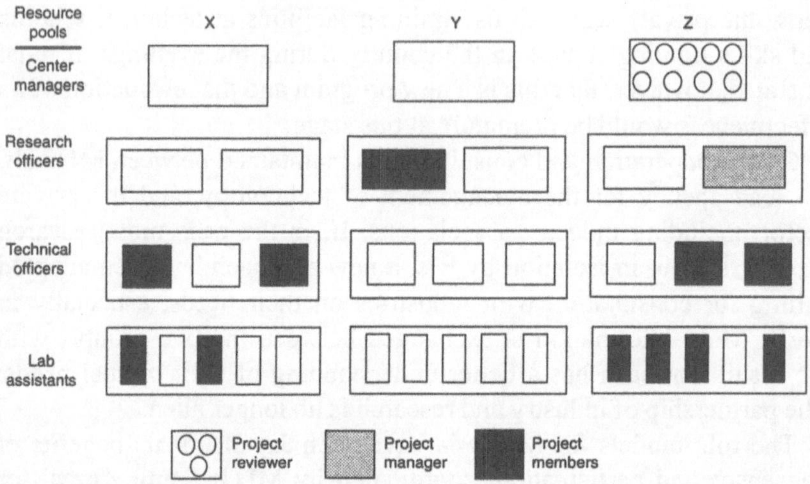

In order to serve clients effectively and manage projects efficiently, the project structure must be strengthened vis-a-vis the line structure so that project commitments take priority.

to engage in research and incremental innovations within their existing business. But this type of organization lacks strategic responsiveness and adaptability. There is no place for new ideas that do not fit into the current divisional program. Vision is therefore short-term.

The ability of the organization to be responsive, flexible, and adaptable is crucial to the success of the desired partnership with industry. A new strategy and solution is the use of project efforts and project management systems. SIRIM has just adopted this kind of structure, graphic representation of which is shown in Figure 8.1. The number of projects increases or decreases with changing diversification strategies and in response to the need for new or improved products or services. Therefore, a variety of interdisciplinary personnel is needed; personnel are drawn from across functional areas and are assembled into a project team under the projects manager. When the project is completed, the personnel return to their original functional areas or are transferred to other projects.

For successful implementation of the innovation process at the commercialization stage, a business and enterprise marketing unit that complements this structure is also desirable.

Implications for Structural Changes

What implications arise from this new structure of operation in the organization? A blend of individual behaviors and specialized activities is required. But because they are reciprocally dependent, these activities must be linked together and integrated into a harmonious system. Thus great skill in coordinating various projects and managing personnel is required. Conflict management, especially in issues pertaining to authority between the project managers and divisional managers, is likely to ensue, which requires skills in human relations management.

Research personnel must also acquire new knowledge on the administrative side. Business orientation and financial accountability become a new focus for administrators. As project managers, they are not line managers alone. They must be trained in conflict resolution, staff motivation, and communication with clients, and learn to manage multidisciplinary projects and deal with contractual issues. The summary of the project management processes shown in Figure 8.2 on page 192 reflects the new skills requirement.

Policy Implications for MOT

As program strategies change in response to industry needs, suitable policy prescriptions to facilitate changes in the management of technology are also required. Recruitment policies, compensation systems, training incentives, rules and regulations, and accounting systems all need review. Just as the response to industry should be quick, turnaround time for support facilities within the organization likewise require improvement. The organization cannot work in its traditional/classical structure if strategies require quick response and a flexible management structure. Goals today are moving targets and are not always clearly definable: a modern concept of structure is required in order to produce creative outputs.

The degree to which industry makes use of services offered by government-owned RIs, universities, or departments that handle tech-

Figure 8.2. **Summary of Project Management Responsibilities**

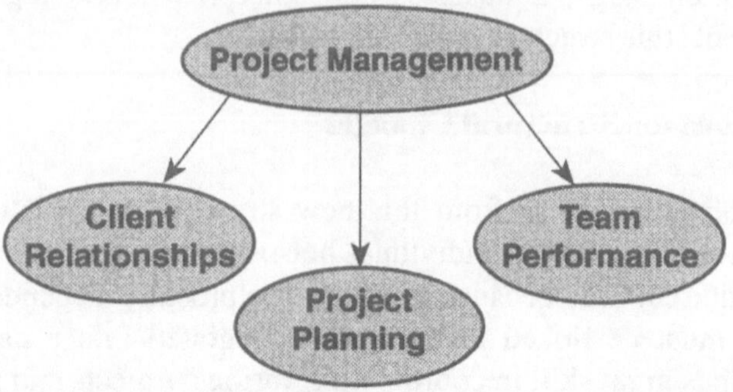

- Business development
- Proposal preparation -contractual, commercial, planning and selling document
- Contractual and working relationships

- Understanding client needs
- Requirements analysis, task breakdown, estimating
- Work program and resource planning
- Staff commitments
- Contractual and commercial offer

- Team management and leadership
- Communications: with client, with team
- Monitoring and control

nology is dependent on its perception of the capability of the organization, the services that are provided, and on its ability to meet the demands both operationally and at the strategic level. This partnership between industry and government is currently working well because of the change in policy in the work relationship with the private sector through the Malaysia Incorporated Concepts introduced in the early 1980s. The basis for trust between the two parties has been molded. Regular dialogue and consultation to review industry's needs have aided the development of programs in technology management. For example, both fiscal and financial incentives for industry to acquire technology and promote R&D were in fact proposed by industry itself.

Conclusion

Articulation of research policies and priorities and explicit government support is essential for the success of human resources programs. Human resource development through the training and retraining of sufficient numbers of researchers is important to meet the growing demands and needs of the rapidly changing industrial sector.

Success in MOT, however, not only requires government to formulate a high-priority program of strengthening R&D capabilities; it also requires the follow-up: a program of mutually reinforcing policies and the provision of the needed infrastructure. The research institutes and the government need to be concerned with improving the institutional structure and procedures involved in formulating, screening, and prioritizing R&D projects; the nationwide coordination and pooling of R&D efforts; and the efficient dissemination of information and effective diffusion of technological applications.

9

Management of Technology: The Indonesian Case

Cornelis P. F. Luhulima

Background

The Indonesian government considers technology to be a vital factor in the development of the country's industrial sector. It becomes a means to accelerate the national development program, while simultaneously being an object of the program. National laws are enacted and presidential decrees and government regulations at the ministerial level issued to direct the implementation of technology as well as the development of technology itself. The establishment of the Office of the Minister of State for Research in 1973 (who became the minister of state for research and technology in 1978) was aimed at coordinating and further promoting the role of S&T in national development.

In addition to increasing the number of government research institutions, the Indonesian government has coordinated research activities. The Office of the Minister for Research and Technology is also instrumental in reducing the duplication of research efforts in order that research activities can be directed toward government priorities decided upon at the sessions formulating each five-year plan (PELITA), one of which has been in effect since 1969. In the process, the government has felt the need to diversify research activities. The universities, for instance, are conducting research activities of a more fundamental character; the Indonesian Institute of Sciences (LIPI) is conducting policy-oriented research, whereas the Agency for Technology Assessment and Application (BPPT), the National Institute for Space Research (LAPAN), the National Institute for Atomic Energy (BATAN), and the National Coordinating Agency for Surveying and Mapping (BAKOSURTANAL) are conducting research of a more applied character. The government has increased the number of researchers by way of educational and training programs, which are also designed to increase the capabilities of Indonesian researchers.

The use of modern technology to exploit national resources and to manufacture goods to meet the people's basic needs has been extensively promoted. The government has determined that existing domestic technological capabilities are not sufficient to fulfill the requirements and to meet the demands of the industrial sector, which is expected to play a prominent role in the economic development of the nation. Hence, the government greatly emphasizes the acquisition of foreign technology to accelerate the industrialization process and the process of production optimization.

Indonesia's S&T development in the first Long-Term Development Plan (1969–94) has been conducted by way of two development activities. First is the transfer of technology, particularly in the agricultural and industrial sectors. This includes activities by the government and the private sector such as technology import, utilization of new technologies, industrial construction, creation of infrastructure, and on-the-job training. The second way of developing S&T is via R&D activities in government institutions and universities and by increasing the quality of the educational system and the number of students. S&T activities in the fields of agriculture and industry are closely linked to economic, financial, industrial, and commercial policies, while S&T development by way of R&D is closely linked to technology policies, education, and the policies of technical ministries conducting in-house R&D to meet their specific needs.

In economic and financial policy, the most significant features are increasing openness to foreign investment, and a more open economic system, both of which have been promoted since the start of the Long-Term Development Plan (the first and second five-year plans). As a consequence, foreign investment in the industrial sector became progressively greater, although limited to import substitution. These initial foreign investment projects, accompanied by the introduction of new machinery, production techniques, and management, have started the process of familiarizing Indonesian technicians and producers with new technologies. Most striking is the mastery of new design in the fertilizer, cement, and construction industries by Indonesian technicians.

Technological development in the agricultural sector was accomplished by the utilization of relatively modern agricultural systems, such as seeds, fertilizers, and pesticides of superior quality, improvement of irrigation systems, and improved agricultural production techniques. The speed with which rice production increased points to the successful application of technology in agriculture. R&D in biotechnology is also being promoted to incrementally increase food production.

The drop in oil prices in the early 1980s forced Indonesia to redesign its economic policies. The focus is now on export orientation. This has been a period of accelerated industrial development on account of deregulation and "debureaucratization," changes that support a more dynamic economic system. A by-product of the acceleration of industrial activity has been the acceleration of technology transfers.

To further promote R&D in the private sector, the government embarked upon a program of tax reductions for companies conducting R&D and technology training to develop the technical capabilities of employees. Apart from this, the government has promoted the establishment of venture capital companies, hoping to persuade industry to make new investments incorporating R&D activities.

R&D activities for the development of science and technology have received strong government support. The government considers S&T to be the backbone of national development in the years ahead. This is obvious from the relatively lavish funds that the government allocates for improvement of the educational system, although a greater part of the funding is dedicated to primary education. This is done on the assumption that well-trained human resources are essential to the development of a nation's S&T capabilities.

Overview of Indonesia's S&T Policy

Indonesia's S&T development is pursued on the basis of five basic principles: (1) the commitment to education and training in the whole spectrum of S&T is requisite to the needs of development and nation building; (2) there must be a realistic and consistent conceptualization of the nature of society, and the technologies that are needed to meet the needs of such a society and to provide solutions to the pertinent issues that such a society encounters; (3) technologies can only be transferred, adapted, and developed in a concrete manner, that is, if they are applied to concrete problems; (4) no country can be a perpetual net importer of technology, so each country will thus have to develop its own technologies; and (5) a country will have to protect the development of its technological capabilities until such time as it reaches international competitiveness.

These principles are to be implemented in four not necessarily successive stages. The first and most fundamental phase is the use of existing technologies through licensed production for value-added processes in the assembly and manufacture of products already available both domestically and internationally. In this phase, manufacturing and management technologies are used to transform raw materials and intermediate products into finished goods of higher value. Progressive manufacturing programs will have to be designed to assist in the systematic transfer of licensed technologies. In so doing, better manufacturing skills as well as better organization and management capabilities, work discipline, and quality standards are inculcated and enforced.

The second phase of technology transformation is the integration of existing technologies into the design and manufacture of entirely new products. Technologies are used to create new blueprints and designs, which means the addition of the element of creation. Apart from the design capability, skills are gained in the integration and optimization of components into new systems and in the capability to select the most effective design for components of the new system. The development of these skills also enhances the opportunity to select from the entire range of globally available technologies the ones that are most suitable for the design and manufacture of the new product.

The third phase is technology development. Existing technologies are improved upon and new ones are developed in the endeavor to design and manufacture future products. At this phase there is the motivation to create completely new technologies. It is the phase of innovations, of the creation of technologies for components to be integrated into products that are most advanced in their markets. Indonesia would be wise to plan on embarking on this phase lest the country lose the advancements already achieved in the first and second stages of the technology development policy.

The fourth phase is large-scale basic research to defend the technological superiority already achieved. This stage is not as central to the transformation of developing countries as are the first three stages.

In all phases the principle of active involvement in manufacturing and production processes is absolutely vital. Technological expertise is gained and improved only as long as it is actively practiced, by way of consistently executed programs of designing and manufacturing technologically viable and economically feasible products (Habibie 1983).

The first of the technology transformation processes is what in the literature of MOT is called the *imitation phase* (Schewe 1992). Schewe estimates that imitation costs are only 60 percent of the innovation cost of a technology. Moreover, the imitator can achieve further cost advantages by learning from the errors of the innovator. However, the risk an imitator faces is that he enters the market at a later time, which may affect the product life cycle. Moreover, successful imitation management should have at its disposal not only technology, intelligence, and production potentials, but also marketing potentials to overcome the existing barriers to imitation. Successfully transferred technology is by definition well-managed technology.

The consequence of the second and third phase of the transformation process is that to secure the ability to compete is not only to improve the production and manufacturing technologies but to develop new products and markets as well. And this compulsion to innovate will result in higher economic risks for the new product. These aspects have been incorporated into the three stages of technology transformation.

Attempts to transfer sophisticated technologies in order to increase production must fulfill two basic conditions. First, the technology transfer must implement progressive manufacturing plans, that is, it must be possible to decompose the production programs into steps enabling the "progressive penetration" of the technology being applied. The second basic condition to be met is that the products or product groups manufactured must be competitive in the market. For this to happen, the scale of production must be optimized, and the quality of production as well as that of after-sales service must be reliable. On the basis of this policy, "bridging the management-technology interface with a scope ranging from invention/creation to customer/service" is guaranteed (Baker 1992, 629). This "bridging" process plays a critical role "in designing the strategies and imparting the skills and attitudes to engineers and managers that they will need in a future technology dominated economy." Baker interprets the U.S. National Research Council definition of MOT as a "business strategy requiring a sequence of steps in the process necessary to achieve long-term strategically determined operational objectives" (ibid. 629).

The product group that constitutes the best vehicle for Indonesia is the entire equipment sector: the aircraft industry, the maritime and shipbuilding industries, and the automotive and rolling-stock industries. Electronics and telecommunications are the next logical categories for technology transfer. The energy industry, including the production of turbines, boilers, generators, and heat exchangers, and energy transport and transmission equipment are other vehicles for transformation. There is also a continuous need to improve the engineering industry and the agricultural equipment industry, and to develop the domestic defense industry.

With the growth of these industries, other sectors of Indonesia's economy will also expand through many backward and forward linkages: road construction, housing, food production, pharmaceuticals, and various kinds of services. Industries that at first would not be appropriate vehicles, such as the production of engines, propellers, undercarriages, and hydraulic systems, would become economical as a consequence of the creation of markets. All these industries can be regarded as the "invisible" ninth vehicle.

In the endeavors to improve Indonesia's technological capabilities in these sectors, most of the first-phase and second-phase development programs are conducted by the Center for the Development of Research, Science, and Technology (PUSPIPTEK) in Jakarta, and other government R&D institutions. When completed, PUSPIPTEK'S facilities will include a construction-testing laboratory; aerodynamics, gas dynamics, and vibrations laboratories; a thermodynamics and propulsion laboratory; a process technology laboratory; a physics laboratory; an electronics laboratory; a chemistry laboratory; a calibration and instrumentation laboratory; an energy laboratory; a metallurgical laboratory; and a multipurpose research reactor and supporting laboratories.

R&D Trajectory

For R&D institutions, the four stages of technology transformation are embedded in four essential components as developed by the Asian and

Pacific Centre for Transfer of Technology (APCTT) of the United Nations' Economic and Social Commission for Asia and Pacific (ESCAP) and the Centre for Analysis of Science and Technology (PAPIPTEK) of LIPI; these are: *orgaware* (institutional framework); *inforware* (information/data); *humanware* (capabilities); and *technoware* (facilities). In any resource transformation, all four components of technology are required simultaneously. No transformation can take place in the absence of any of the components. Technoware is the core of any transformation facility. Humanware is the key element of any transformation, and it develops, installs, and operates, whereas inforware is generated and simultaneously used by humanware for decision making and operation. Orgaware acquires and controls inforware, humanware, and technoware to effectuate the transformation. The handling of technoware can proceed efficiently and effectively if humanware possesses the relevant knowledge/information/inforware to imitate or develop a new technology as a value-added process. Thus the understanding of inforware is a process of societal/national education by way of the abstract comprehension of a technology. (See Pusat Analisa 1987; this report was prepared by a technical team at the Asian and Pacific Centre for the Transfer of Technology in Bangalore, India, and adopted by the Indonesian Institute of Sciences.)

Humanware in this context is the skill or expertise to use and develop a technology, which necessarily means the successful comprehension and application of the technology. However, it should be understood from the very start that inforware, technoware, and humanware will not function efficiently and effectively if the orgaware does not function equally well. Hence, every process of the transfer of a technology is highly dependent not only upon the technical knowledge but on the environment and the capability to harmonize technoware, inforware, and humanware, which is essentially a management factor.

The level of sophistication of the four components of technology can increase by way of:

- increased operational complexities;
- increased value of documentation;
- increased requirements of individual skill and experiences;
- increased interaction and dynamism.

Industrial Trajectory

Most of the manufacturing industries in the private sector are in the first stage of technology development. In the government enterprises, the stage of development is generally higher, particularly for those in the strategic industry group, which is technically responsible for the creation and management of strategic industries that devote their technology, productivity, and efficiency to support national development and ensure self-reliance in national defense and security. There are nine state-owned companies that are classified as strategic industries and one former production unit that has been converted into a corporate entity. They are:

- PT Barata Indonesia—produces heavy equipment, construction materials, and machinery, and delivers engineering services;
- PT Boma Bisma Indra—produces diesel engines, construction equipment, and machinery and delivers engineering services;
- Perum Dahana—a state-owned corporation engaged in the production of explosives;
- PT INKA—produces railroad cars and accessories;
- PT IPTN—produces aircraft and weapons systems;
- PT Krakatau Steel—manufactures integrated steel products;
- Production unit LEN-BPIS, the predecessor of PT LEN industry—a professional electronics and components industry;
- PT PAL INDONESIA—nautical and engineering ventures; produces marine vessels;
- PT PINDAD—a weapons and munitions industry.

These enterprises are all at the second stage of technology development; they function as vehicles for technological and industrial transformation.

The output of these industries has to fulfill two specific requirements: first, to manufacture the products and, more important, to manufacture them competitively, a progressive manufacturing plan must be followed; second, the products have been manufactured on the basis of an economy of scale. To fulfill the second requirement, products will have to be publicly procured, mostly to support the nation's economic infrastructure (railroads, ships, airplanes, telecommunications, etc.).

The following other institutions are engaged in technology transfer:

- the Investment Coordinating Board (BKPM), which has the task of formulating capital investment policies and evaluating their implementation;
- the Agency for the Assessment and Application of Technology (BPPT), which has the task of monitoring and controlling technology transfer and conducting studies on the impact of industrial technology on other aspects of national development activities and on social life in general;
- research institutions of the Indonesian Institute of Sciences (LIPI) and others, assigned to develop and adapt foreign technologies to support various industrial activities;
- universities that disseminate technological know-how through education and training;
- industries and associations of industries.

Presently, foreign technologies being applied in Indonesia have to use as many local components and materials as possible; the condition is imposed to support government policies to increase the use of local products. The impact of foreign technology on indigenous S&T development is expected to enhance the ability to understand new techniques, designs, and production processes; integrate existing indigenous technologies with new production systems; and enhance the ability to develop new technologies.

In the process of technology transfer, it is compulsory for foreign investors to provide facilities for the training of Indonesian personnel. Training, whether locally or abroad, should enhance the technology selection, handling, and development of technology and the gradual replacement of foreign staff with local personnel.

As the private sector deepens its technology involvement, there are problems inherent in the transfer of technology from a proprietor to a receiver. This is primarily caused by the two venturing parties themselves. Not all technology proprietors are interested in a long-term association with the receivers. Similarly, not all technology receivers are interested in developing an in-house capability beyond manufacturing.

Generally speaking, the biggest incentive for proprietors to transfer their technology is the opportunity to widen their markets overseas, to increase their sales volumes, and to raise R&D funds to further develop their technologies. One particular dimension in furthering development of technologies is the design of cooperation programs in R&D with the technology receiver. As receivers make better use of the transferred technology, they become stronger and more skilled not only technologically but also financially. Through increased purchases of goods and services from the technology transferors, the technology receivers can contribute to the funding of the transferors' R&D.

Another factor is that the property rights of the technology owners must be protected. The existence of a succinct body of laws and regulations that sufficiently warrants intellectual property and patent rights protection is an important preparation that a technology receiver has to make.

An important factor for the technology transferors is that the transfer of their technology does not lead to job losses—hence the necessity of establishing a division of labor between technology transferors and receivers. This division in value-added processes is accompanied by the sharing of the market. The technology transferors must be guaranteed that the technology receivers will not compete in the transferors' market, while, similarly, the technology receivers must be given sole marketing rights in the appropriate domestic area and the surrounding areas.

The technology receivers have specific interests as well. First, they must be guaranteed that the technology they receive, under various schemes comprising the division of labor and market sharing, is state-of-the-art technology. Second, the technology receivers should be guaranteed that they will always obtain the latest information and the most recent technology. Third, the technology receivers' economic and human resources must be involved in the further development of these technologies. Fourth, a successful technology transfer should be preceded by the training of the labor force. In the majority of cases, the labor force has just completed formal training and as a consequence lacks practical experience. Moreover, the value-added processes that the technology receivers undertake lack the integration usually present in the modern organization of the technology transferor.

On account of these factors, the technology receivers have to undertake additional investments in infrastructure, plant facilities, education and training, and systems and work procedures in their endeavors to transform themselves into technology carriers (Samadikun 1992, 5ff).

According to one study (Thee Kian Wee 1990), technology transfer and product adaptation in the manufacturing plants of a number of transnational corporations (TNCs) in the pharmaceutical, food, chemical, and automotive industries in Indonesia consist largely of minor modifications to reflect differences in taste (greater preference for sweetness in biscuits, for example) in available ingredients (ice cream flavored with local fruits), and in weather conditions (paints). Product adaptation in the case of pharmaceuticals is nonexistent, since TNCs are greatly concerned with the maintenance of product quality and the reputation of their products.

In the case of process adaptations, most of the firms surveyed—in particular, those involved in joint ventures—were only willing to introduce modification peripherally, such as in the packaging, sorting, and handling of materials, where more manual labor was used than in the TNC's home country or in their subsidiaries in other industrial countries and where the level of mechanization or automatization was usually considerably higher. In the stage of the "core" production, however, generally no significant process modifications were made, primarily due to concerns about maintaining quality.

None of the companies had developed an indigenous R&D facility to conduct basic and applied research. The facilities that they established are used mostly for such activities as product testing, quality control, introducing minor product modifications, or solving minor technical problems in the production process. These small laboratories are unlikely to develop into full-fledged R&D facilities because of the lack of an adequate supply of skilled Indonesian personnel and the relatively small size of the local market. A regional market has not yet taken shape as a consequence of the establishment of the ASEAN Free Trade Area (AFTA).

Definition of the Management of Technology Challenge

The current debate on economic development in Indonesia reflects two seemingly different strategies. The traditional approach is that economic growth should be promoted primarily through exogenous technological change. The attention given to technology development and the accumulation of human capital is relatively small. The major focus of this policy is macroeconomic stability, particularly monetary stability. It is a policy of balanced budgets, low inflation, competitive interest rates, and free-market orientation on the basis of comparative advantages of labor-intensive industries. The newer approach is associated with value-added economic growth—growth that is primarily promoted through endogenous technology development supported by the accumulation of human capital, both through formal education and the informal development of skills in the workplace. Technology is thus embodied in human capital.

The new economic development strategy for the next twenty-five years emphasizes that the rate of physical capital accumulation is a function of the level of human capital. Thus the accumulation of human capital has a higher priority than the acquisition of physical capital. It is the approach of "threshold human capital"; below the threshold, a country will be trapped at a standstill; above it, high growth rates can be achieved. Hence, a state's capability in technology development and the accumulation of human capital are determining factors in a nation's development.

The core of the new strategy is to transform human resources to become more productive in certain categories of production in order to support the mastery and development of technology. The new approach also focuses on growth effects, that is, fundamental changes in the parameters, which will alter significantly the blueprint for long-term economic growth. It is characterized by "advocacy policies," which are explicit in determining the objectives of industrialization, human resource development, and the promotion of technology adaptation, particularly by the government, and which apply specific market distortions to guide the economy toward certain objectives, building up domestic production not necessarily in tune with the international division of labor. In a number of cases the policy of promoting certain strategic industries or sectors will be applied. Thus the policy of industrial deepening is promoted by a range of policies for the development of technology and human resources. Technology policies comprise the promotion of technology transfer, of R&D, training of experts, and directing foreign investment that sustains technology transfers.

The policy of human resources development comprises the development of institutions for personnel training, with emphasis on the development of knowledge and basic skills with sufficient levels of flexibility as well as the ability to read market signals. Most important for Indonesia from now on is the incremental increase in the number of polytechnic institutions and their graduates.

However, it is widely acknowledged that the second approach cannot be pursued without sustained macroeconomic stability, which was the focus of the first twenty-five years of the New Order government. It is also realized that the new approach, with its emphasis on value-added and technology leaps, does not promise high economic growth in the short term. It is highly possible that economic growth in the short term will be lower than before, because funds will be redirected toward the mastery of technology and the development of strategic industries by way of protection and subsidies. Technology leaps require sufficient highly skilled human resources (critical mass), the ability and vision to determine strategic industries, a strong private sector that will translate the second approach in a practical way, and sufficient financial resources.

This new approach will, however, focus primarily on the economic theories of industrial organization and will renew the argument for industrial protection. The free-market principles to which the first approach strongly adhered did not give ample opportunity for R&D to develop, since in the free-market system enterprises were reluctant to invest significantly in R&D. Government intervention through protective tariffs, subsidies, and protection of patents will arguably encourage Indonesian enterprises to invest in R&D, thus leading to the development of new technologies.

This new approach to Indonesia's economic development will presumably augment interest in technology development and, as a corollary, in the management of technology.

Organization of MOT

Public Sector Strategies

Although the four technology transformation processes have been adopted as the principles of MOT, a system for organizing MOT training is presently not available nationally. Such training is conducted on a sectoral basis.

Human resource development in the agricultural sector is conducted by the Agency for Agricultural Research and Development (AARD) of the Ministry of Agriculture and comprises the following: increasing the qualifications and capabilities of research staff in the agricultural sector, and the number of research staff; short-term training abroad to develop skills and knowledge using endogenous resources or donors and loans from the World Bank and the United States Agency for International Development (USAID); sending research staff for postgraduate training, within the country and abroad; training for technicians in special skill development at AARD; encouraging research staff to participate in seminars and workshops to develop their knowledge and to gain experience, both domestically and abroad.

The other ministry that carries out a series of programs to develop technical skills and entrepreneurship to support the manufacturing sector is the Ministry of Industry. Its activities include the following: making estimates of the need for national personnel including experts and managers; conducting supplementary training to develop skilled personnel based on the stage of development of manufacturing technology; encouraging the establishment and development of top- and middle-level management courses in cooperation with third parties and universities; encouraging special skill training and apprenticeships by manufacturing industries (APCTT 1990, 61–62).

However, human resource development programs in both institutions have not incorporated MOT. That was reserved for the Indonesian Institute of Sciences (LIPI), which, since 1970, prior to the formulation of the four stages of technology development, had organized two- and three-week R&D management workshops to enhance the capabilities of researchers and managers in managing R&D projects and institutions in tune with the needs and dynamics of national economic development. While this program has been managed from within LIPI, it has received active support and assistance from many of Indonesia's ministries, universities, and nonministerial organizations. Indonesian R&D managers have participated in management workshops in both Indonesia and the United States. The R&D management program has been financed by the Indonesian government with the assistance of UNESCO (the United Nations Educational, Scientific, and Cultural Organization) in 1970, Inveresk Research International (IRI) of Scotland and between 1975 and 1987, the Denver Research Institute. Workshops carried out under the LIPI program have provided participants with exposure to and training in state-of-the-art R&D management practices. They have covered such management areas as R&D project management, R&D institutional management, R&D project and program evaluation, technoeconomic feasibility studies, R&D monitoring and control, scientific and technical information for

R&D institutes, R&D institute-client relationships, training of R&D trainers, and R&D management consultancy.

During the 1980s, these workshops focused on preparing a cadre of Indonesian R&D management trainers and consultants to perpetuate the development and implementation of improved management techniques (Black and Luhulima 1984). The 1983 Workshop on R&D Management Consultancy Services produced an anthology of diverse consultative essays prepared by the participants on the topics of R&D project selection and funding, the role of clients in R&D, manpower development (as it was then called) and productivity, the flow of communications and information in R&D, and R&D institute evaluation. The project was discontinued in 1988. This project of R&D management training was the only cross-sectoral and cross-institutional training in the country in what is now called management of technology.

Early in 1993 the Agency for Technology Assessment and Development (BPPT) in cooperation with the Korean Institute of Sciences and Technology (KIST) organized another R&D management course focusing on the project process management system and the related concepts of R&D management, R&D/innovation process and strategy, long-range planning for R&D project management, R&D project process management, and new directions and special topics in R&D/technology management. The brief considerations given to technology covered technology transfer and diffusion, technology forecasting, and technology assessment. The course met one day a week for six hours over three months. The participants, however, were limited to BPPT and LIPI personnel.

Academic Strategies and Programs

Bandung Technological Institute now offers an M.B.A. in technology. The program is primarily meant to create a business outlook among engineers, particularly with a view to the marketing of technology, so that the engineers have the capability to cope with the fast-changing pace of industrial development. The program is not meant to develop skills to manage technology.

PT Telekomunikasi Indonesia under the presidency of Cacuk Sudarijanto has established a school for management as well. The program is run in cooperation with the Manila-based Asian Institute of Management and caters to the general public as well as to PT

Telekomunikasi. Here again the objective is to prepare its personnel and the public in the area of marketing telecommunications technology, rather than in MOT.

Since the 1980s, M.B.A. programs have been offered by private institutions in Indonesia. The most famous among them are Institut Manajemen Prasetya Mulya (IMPM), established in 1982, Institut Pengembangan Manajemen Indonesia (IPMI), begun in 1984, and Institut Pendidikan dan Pembinaan Manajemen (IPPM), which was established in 1985. However, none of these institutions offer MOT programs.

Private Sector Organization for MOT Training

In most private industries the necessary qualifications for managers and technical personnel are primarily acquired through short-term on-the-job training and additional training at TNC plants in the home country or in their overseas plants located in other industrialized countries. Additional training is also provided to local employees by expatriate managers and technical experts assigned by a TNC to work in a joint venture, or with a national company that has a technical licensing agreement with a TNC. Specific MOT training is not yet under way.

Mechanism for Interaction

The technology infrastructure of a country comprises academia or universities, R&D institutions, and engineering and production units. The term *innovation triangle* refers to the interaction among the three pillars of the infrastructure in research as well as development work. This interaction could be referred to as the industrial research triangle and has a crucial role in advancing science and technology.

Academia and R&D institutions supply qualified human resources and technical know-how, while the engineering and production units supply both the means and the opportunity to exploit the R&D output for the production of socially demanded products at affordable prices.

In Indonesia there is a considerable lack of proper interface among the three units. The technology developed by the R&D institutions is generally not ready to be transferred to industry, or industry is not prepared to absorb and apply the R&D output due to technical and economic considerations (Hermawati 1993).

It is the effective linkage among the three units that will determine the country's ability to develop endogenous technology, to effectively use the transferred technology, and to develop indigenous R&D capabilities, which is a sine qua non for successful industrial development. The development of a strongly constructive relation among the three flows from such factors as funds, technoware, humanware, inforware, and orgaware.

The triangle relationship is best illustrated by examining the triangular cooperation among an electronic public enterprise (PT LEN) as the production unit, TELKOMA–LIPI as the R&D unit, and academia (ITB). Here the strength of the linkage is primarily determined by the magnitude of the funds flowing from PT LEN to TELKOMA or to ITB. This flow of funds is the most critical factor in developing the link. Joint research is the main link among the three.

Summary Profile

There still is little interest in running MOT programs in the public, academic, and private sectors. With a view to the second twenty-five-year development plan, which began in April 1994, a conducive atmosphere has been created to develop MOT. Hence, if Indonesia does not want to be left behind in the race for the technology capability to pursue national objectives, initiatives must be taken soon to start training technology managers.

The Prevailing Influence on MOT

Historically, scientific development in Indonesia was promoted by the Dutch during the colonial period and continued after independence in the universities and research institutes under the Indonesian Council for Science. However, it was only in the late 1960s and early 1970s that technology became a central topic in R&D in Indonesia. It was during this period that technology transfer, adaptation, and development were fostered through foreign direct investments and licensing agreements, which became the core of Indonesia's economic development.

American influence was very strong in the oil industry and spread swiftly throughout the country, particularly in the early 1970s. The Japanese then entered the manufacturing sector and spread their influence at major industrial sites. It was by way of joint ventures and licensing agreements that technology developed, as did MOT, although on a very limited scale.

In research institutions and universities, foreign influence on S&T development came primarily from Europe, the United States, and Japan. The United States Agency for International Development (USAID), the Science Policy Research Unit (SPRU) of Sussex University in England, Aachen Technological Institute in Germany, Organization for Economic Cooperation and Development (OECD), the National Science Foundation of the United States, the United Nations' APCTT-ESCAP, and KIST were the main institutes influencing S&T development, both through expatriates and through Indonesians studying abroad. However, it should be emphasized that awareness of the need for technology and for the development of human resources was essentially internal, brought on by economic and technical competition for international markets. The recognition that in R&D, management and thus MOT are highly important managerial issues grew from such awareness.

Policy Implications

The fact that R&D management became and still is an important managerial issue indicates that in various circles in Indonesia it is recognized that new initiatives have to be taken to reintroduce MOT in a systemic way and, hence, on an integrated basis. MOT is essential for the development of Indonesia's technology infrastructure in the universities, R&D institutions, and industry, which is still lacking in the country. MOT will contribute considerably to the interface among the three pillars of the infrastructure. External pressure in the form of trade constraints, concern for human rights, support of democracy, "good governance," property rights, environmental concerns, and market mechanisms have forced Indonesian policy makers, businesspeople, and researchers and academics to rethink their approach. Moreover, competition has fiercely increased. Indonesia can be recognized and respected internationally only if it upgrades its capabilities for developing technology to produce new products economically, develops its human resources to the highest level possible, and develops MOT. It is only by creating such an environment that Indonesia will be able to live up to the demands of this competitive world. And the development of MOT is a very crucial dimension of the endeavor to survive in such a world.

References

APCTT (Asian and Pacific Centre for the Transfer of Technology). 1990. *Technology Policies and Planning Indonesia*. 2d ed. New York: United Nations.

Baker, Merl. 1992. "Contemporary Views of Management of Technology and Organizational Implications." In *Management of Technology III: The Key to Global Competitiveness*, ed. Tarek M. Khalik and Bulent A. Bayraktar. Proceedings of the Third International Conference on Management of Technology, 17–21 February, Miami.

Black, Ronald P., and A. S. Luhulima, eds. 1984. *Anthology on Indonesian R&D Management*. Jakarta: Lembaga Ilmu Pengetahuan Indonesia.

Habibie, B. J. 1983. "Some Thoughts Concerning a Strategy for the Industrial Transformation of a Developing Country." Address delivered to the Deutsche Gesellschaft Für Luft- und Raumfahrt, 14 June, Bonn.

Habibie, B. J. 1993. *Pembanqunan Berdasarkan Nilai Tambah dengan Orientasi Teknologi dan Industri*. Working paper no. 1, Seri Dialog Pembangunan 1993. Jakarta: Center for Information and Development Studies (CIDES).

Hermawati, Wati. 1993. "A Methodology for Assessing the Strength of Industrial Research Triangle Linkages." Master of Business Administration thesis, Asian Institute of Technology, Bangkok, Thailand.

Pusat Analisa Perkembangan Ilmu Pengetahuan dan Teknologi. 1987. *Technology Atlas: An Interim Report*. Jakarta: Indonesian Institute of Sciences and Asian and Pacific Centre for Transfer of Technology (ESCAP).

Samadikun, Samaun. 1992. "Technology Transfer and Industrialization. Interaction for Progress: Modernization of Vietnam and ASEAN Cooperation." Paper presented at the International Symposium on Interaction for Progress: Modernization of Vietnam and ASEAN Cooperation, 21–23 May, Vietnam.

Schewe, Gerhard. 1992. "The Imitation as an Option of the Management of Technology." In *Management of Technology III: The Key to Global Competitiveness*, ed. Tarek M. Khalik and Bulent A. Bayraktar. Proceedings of the Third International Conference on Management of Technology, 17–21 February, Miami.

Thee Kian Wie. 1990. "Indonesia: Technology Transfer in the Manufacturing Industry." In *Technological Challenge in the Asia-Pacific Economy*, eds. Hadi Soesastro and Mari Pangestu. Sydney: Allen & Unwin.

10

Management of Technology in China

Wu Yikang, Dong Guilan, and Sun Mengxin

Background

Profile of Economic Structure, Industrial Organization, and System of S&T Innovation

Historically, China was a country with an agriculture-based economy, and its industrialization process started rather late. By 1949, when New China was founded, the fixed assets accumulated by the industrial sector reached only RMB Y10 billion, and industrial production in terms of value accounted for 12 percent of the gross national income. Industrial employment made up 2 percent of the total labor force.

After the founding of the People's Republic, China started its first five-year economic development plan in 1953. Through forty years of continued efforts, especially with the completion of the Seventh Five-Year Plan (1986–90), there have been tremendous changes in China's economic structure. By the end of 1992, gross domestic production (GDP) reached RMB Y2,393.8 billion. Industrial production value reached RMB Y1,011.6 billion, agricultural production RMB Y580.8 billion, and tertiary industry RMB Y662.3 billion. Table 10.1 breaks down the GDP by industry.

Note: Statistical and gross domestic production information cited in this chapter is from the State Science and Technology Commission, People's Republic of China.

Table 10.1

GDP Share by Industry in 1992 (%)

Primary industry	24.2
Manufacturing industry	42.3
Tertiary industry	27.7

Source: China Statistical Yearbook 1992.

The year 1992 witnessed rapid growth in international trade. China's total trade volume ranked eleventh in the world. The total volume of exports was valued at U.S.$85 billion, an 18.2-percent increase over 1991, whereas imports were valued at U.S.$80.6 billion, a 26.4-percent increase over 1991. The year 1992 also saw a favorable balance of payment in international trade, with a trade surplus of U.S.$4.4 billion. With a stable economic, political, and social situation, and a continually improving investment environment, the international balance of payment should become more and more favorable.

In 1992, per capita income increased. The average annual income of city residents reached RMB Y1,826, a net increase of 8.8 percent over 1991.

Within the national economy, the growth of major sectors in 1992 was kept in appropriate proportion according to the state plan. The industrial structure was readjusted and improved. Production at large and medium-size state-owned enterprises increased by 15.5 percent. In terms of increased production value, state-owned enterprises accounted for 43.2 percent.

The economic system reform that was led by rural reform in 1978, brought dramatic changes in the operation of China's economy. In 1978 the industrial production value of state-owned enterprises made up as much as 78 percent of total industrial production, and the remaining 22 percent was nearly all collectively owned enterprises, whose operation mechanism was almost entirely indistinguishable from that of the state-owned companies. However, since the reform in 1978, there has been marked development of exclusively foreign-owned companies, as well of privately owned firms, along with sharp expansion of township enterprises and joint ventures with foreign companies. In 1992, the production value of township enterprises constituted over 30 percent of the total increase in industrial production, and the aggregated increased production value of joint ventures, foreign-owned companies, privately owned firms, and township enterprises made up 61 percent.

The above-mentioned non-state-owned enterprises have been operating as independent commodity producers with flexible market mechanisms that are subject to practically no government intervention. Meanwhile, state-owned enterprises have been undergoing a process of transformation with the goal of building up a market-oriented mechanism.

Now China is in a transition toward a socialist market economy. Within the state-owned enterprises the share of centralized planning is decreasing and decentralized planning is becoming far more important. The variety of products assigned for production through centralized planning is on a steady decline, from 120 in 1984 to 65 in 1991. The share of production value created through the process of centralized planning dwindled from over 80 percent in 1988 to 16.2 percent in 1991, while the production value gained through the process of decentralized planning rose to 42.9 percent, with the rest left to the market force to regulate. According to state statistics in 1993, there are 407,989 industrial enterprises in China, employing more than 100 million people.

The Chinese industrial system covers almost all the industries, ranging from light to heavy, from labor-intensive to capital-intensive, and from traditional to new and emerging. There is no doubt that after forty years of continued efforts since the founding of the People's Republic, China has established a fairly complete and well-distributed industrial system.

Meanwhile a fairly complete system for S&T innovation has also been built. This is a pluralistic networking system joining the central government with the local government, the executive agencies with management bodies, and different fields and industries, with government playing a dominant role. Some of these S&T innovators are described below.

1. R&D institutions attached to government, institutes of higher learning, and industrial enterprises are the three major sectors carrying out S&T innovations and the mainstay for S&T development in China. Among them, R&D institutions attached to government have the most renowned scientists and engineers and better research equipment and

have made the greatest investment in R&D. They have produced as much as 60 percent of total R&D results. In 1991, R&D institutions attached to government totaled 5,487, of which 123 were under the jurisdiction of the Chinese Academy of Sciences, 1,070 under control of mission ministries and commissions of the State Council, and 4,222 under the jurisdiction of local governments at and above the county level. Total employees numbered over one million, of which 380,000 were scientists and engineers; R&D expenditure totaled RMB Y16.9 billion. Of all the institutions attached to government, those under the jurisdiction of mission ministries and commissions were the strongest in terms of resources devoted to technology development. They are regarded as the most significant force in technology development, having 61.8 percent of the scientists and engineers and making 65.3 percent of total R&D expenditure.

2. Another important sector in the execution of S&T activities is universities and colleges in China. In 1992 there were 2,230 R&D institutions attached to universities in the areas of physical science, engineering, agriculture, medical science, and teaching. They employed almost half a million scientists and engineers and had an R&D budget of RMB Y1.2 billion.

3. Large and medium-size industrial enterprises comprise the third sector performing S&T activities. In 1992 there were 13,475 such enterprises in China, of which 8,522 had in-house R&D institutions. These institutions are recognized as the major force for technology development, directly oriented to new products, new processes, and technology and equipment updating.

In addition, an increasing number of non-government-owned organizations for technology development, some collectively owned and some privately owned, are now in the midst of R&D management system reform. In 1990 there were 8,523 such organizations employing 170,000 people, of which 110,000 were scientists and engineers. As the reform in China continues, the number of such organizations will greatly increase.

National S&T Policy

In the forty years since the founding of New China in 1949, China has been transformed from a country extremely backward in S&T into one with reasonable S&T capability. The rapid increase in S&T capability played a significant role in China's economic development and laid a solid foundation for the realization of the "four modernizations": agriculture, industry, defense, and science and technology. The development of S&T in China can be divided into the following four periods.

The First Period (1949–66)

Immediately after the founding of New China, the government began to devote much of its attention to developing S&T. By 1955 there were over 840 research institutes, and the total number of S&T personnel reached 400,000. S&T began to take shape and played an active part both in the period of economic recovery (1950–52) and during the First Five-Year Plan (1953–58). In 1956 China's first Long-Range Program for Scientific and Technological Development was drawn up, in which major projects in basic research, applied research, and development were identified. By 1962 most of the objectives in this plan had been fulfilled, five years ahead of schedule, so a Program for Scientific and Technological Development for the years 1963–72 was mapped out, identifying 374 key research and development projects. In October of 1964 China successfully conducted its first nuclear test, which marked a new stage in China's S&T achievements.

The Second Period (1966–76)

The ten-year-long "Cultural Revolution" that erupted in 1966 caused disastrous setbacks to China's S&T development. The whole R&D system was paralyzed: research institutions were disbanded; laboratory equipment, research facilities, and S&T data and information were destroyed; and all major R&D activity ceased.

The Third Period (1977–80)

After the downfall of the Gang of Four, tremendous efforts were made on the part of the Chinese government to restore the country's economic growth. The State Science and Technology Commission of China and its local chapters were reorganized to take full charge of S&T development. The Outline Program for National Scientific and Technological Development for the years 1978–85 was formulated. In particular, 108 research projects were identified as priorities in the areas of agriculture, energy, materials, computers, laser technology, high-energy physics, and genetic engineering.

The Fourth Period (1981–85)

After 1978, when the focus of the government's attention shifted to economic development, there was a growing public awareness of the role that S&T plays in economic growth and social progress. Thus S&T has been given priority and support in the country's development program, which has culminated in the formulation of the guiding principle that "economic development must rely on science and technology, which in turn should be oriented toward the former." This has become a basic policy for all sectors to follow, including industry, agriculture, and enterprises. The focal points in this guiding principle were as follows:

- S&T should advance in coordination with economic growth and social progress, with the promotion of economic growth as its first objective.
- Emphasis should be placed on the R&D of production technology and on the formation of appropriate technological structure.
- Technological capability in industry and agriculture should be strengthened.
- Gradual development of basic research should be maintained.
- S&T advancements from abroad should be regarded as an important way to increase China's own capabilities.

These guiding principles proved effective in practice and have been pursued ever since. For the realization of S&T development in coordination with economic growth and social progress, reform of the S&T management system was initiated in 1985 and has been carried on since then. Included in the reform was the change from an R&D funding system entirely dependent on government to one with several funding sources. R&D institutes were also encouraged to undertake contract and commissioned research for enterprises and were given giving more decision-making power, and scientists, engineers, and other technical personnel were encouraged to accept employment in enterprises in remote or rural areas, or to set up their own technology-based firms.

Through the reform, a close tie between S&T and production was successfully forged, and the initiatives of scientists and engineers were encouraged. It is still going on in depth and scale.

The Challenge of MOT in China

With large numbers of qualified scientists and engineers committed to scientific research, China is quite strong in R&D, and tremendous achievements have been made. However, under a centrally planned economic system, R&D was separated from production. Managers at research institutions were very competent in planning, organizing, and coordinating research projects funded by government, but after the completion of the project, all they had to do was to deliver the results to government. Under this system, they were not expected to transfer the technology into production or to commercialize it. Now China is in transition from a planned economy to a market economy. There is a pressing demand for managers who can transfer and commercialize technology and can also adapt to the norms of a market economy and interface with the international market.

There is a pressing need to transform China's management system and economic mechanisms so that they are compatible with international norms. (For instance, after China's status in the General Agreement on Tariffs and Trade/World Trade Organizations [GATT/WTO] is restored, China will introduce a new accounting system compatible with the internationally accepted one, which will require accountants with specialized training.) A great number of technology managers are needed in China, but the supply is short of the demand. Demand for MOT training, and for the vitally important training of qualified college and university teachers, is growing.

Organization of MOT

Public Sector Strategies

The State Council is the supreme administrative body of the People's Republic of China. It is mandated to administer all affairs, including education, science and technology, public health, physical culture, and family planning. Under the State Council, there are ministries and commissions, of which fifty have set up departments for S&T development or S&T education. The State Science and Technology Commission (SSTC) takes full charge of S&T development in China on behalf of the State Council. The responsibilities of SSTC include studying and formulating national S&T policies, coordinating major S&T projects when their execution involves more than one ministry, and mobilizing S&T resources when troubleshooting is needed for key S&T projects. SSTC is also responsible for allocating and administering the government budget for S&T development and for organizing the evaluation and assessment of major projects of technology importation, equipment importation, and foreign capital inducement. And being a competent authority for international cooperation in S&T, the SSTC coordinates major activities in the field.

Under SSTC are about ten institutions that are engaged in S&T policy studies, management, and training. For example, the National Research Center for Science and Technology for Development is an institution for science policy research and consulting; the Science and Technology Management Staff Training Institute has the responsibility of providing long- and short-term training in S&T management; the China Science and Technology Exchange Center promotes the exchange of information and cooperation between China's scientific community and corresponding nongovernmental S&T organizations abroad; and the Institute of Scientific and Technical Information provides technical services.

The State Planning Commission under the State Council is a coordinating authority for national economic development. Its departments for S&T planning and for social development planning make important decisions regarding resource allocation for S&T and education.

The State Economic and Trade Commission under the State Council is responsible for allocating government financial resources for technology development to commercial enterprises and for organizing trial production of new products, industrial demonstrations, technological updating of enterprises, and the adaptation, improvement, and diffusion of technology imported from abroad.

The State Education Commission is another body under the State Council. It takes full charge of education for the whole country: primary education, secondary education, tertiary education, vocational and technical education, adult education. It also oversees international cooperation and exchanges.

The Office for Talent Import was created in 1985. Its main function is to organize overseas short-term training for technical professionals and to recruit technical professionals from abroad to lecture, conduct joint research, and provide technical consultancy for given projects. Its purpose is to help upgrade the technical and management competence of the Chinese S&T community.

The Ministry of Personnel is an administrative body with numerous functions in human resources development in China. Its main responsibility includes making policies and regulations pertaining to employment, wages, vocational and technical training of the workforce, and continuous education of S&T personnel. In addition, it formulates policies regarding the assessment, promotion, and appointment of professionals in technical positions.

Generally speaking, industry in China is under the jurisdiction of the various mission ministries: the ministries of Agriculture, Machine-Building and Electronics Industry, Metallurgical Industry, Geology and Mineral Resources, Aeronautics and Space Industry, Water Conservancy and Power, Forestry, Light Industry, Textile Industry, Chemical Industry, Railways, Transportation and Communications, Post and Telecommunication, and Energy.

In addition to the department of S&T or of education that each ministry has set up to look after such work as technology development policy, identification of S&T projects, and training of S&T personnel, each ministry has a university or college attached, which trains specialized people in industrial management and in the conducting of R&D in line with the particular field of the ministry.

Following the pattern of administrative setup in the central government, each province, autonomous region, county, and municipality contains corresponding administrative organizations. For example, at the central government level there is the State Science and Technology Commission, and there are S&T committees at the provincial, municipal, and county levels that form an administrative network.

Academic Programs Related to MOT

Universities and Colleges

Universities and colleges are recognized as the major force for MOT training in China. In 1991 there were 1,075 universities and colleges in the country, of which 50 were comprehensive universities and 286

Figure 10.1

Courses Related to MOT in the School of Business Administration at Qinghua University.

Statistics	Accounting Principles
Computer Language	Technological Economy
Control Engineering	Data Structure
Marketing	Business Management
Management System Simulation	Applied Mathematical Studies
Operational Research	Investment
Econometrics	Mathematical Economy
International Technology Transfer and Joint Ventures	Decision Support Systems and Expert Systems
	International Monetary and Banking Systems

were universities of physical science and engineering. It is these 336 universities that perform regular training in MOT. A recent development is that many universities of physical science and engineering have established colleges of business administration focused on providing training in the integrated skills of economic and technology management for people with engineering backgrounds (see Figure 10.1.) In 1991 the number of registered university students totaled 2,043,662, of which 836,872, over 40 percent of the total, were students of science and engineering.

Universities and Colleges for Adult Education

An important component in the education system for technology management is adult higher education, which provides training to government employees, other workers, and farmers. In 1991 there were 1,256 universities and colleges for adult education. Among them, colleges for management specialize in the training of middle-level managers and S&T personnel designated to take management positions. All ministries and commissions under the State Council have their own college for management staff. In 1991, 25,600 people completed training at such colleges.

Industrial Sector MOT Training

Industrial sector MOT training includes training provided by government organizations and academic institutions and in-house training by industrial enterprises. MOT in the industrial sector is defined as the planning, organizing, and supervising of technological research and development, new product and process development, and technological updating and innovation. Its purpose is to ensure rational deployment of technical capability and available resources for rapid expansion of production capability and higher economic return.

Training by Government Organizations

The training of the business administration staff of state-owned enterprises is within the area of government responsibility. MOT training is provided for promising young and middle-aged workers and slated for management positions. As of 1993, selected universities and colleges have sponsored short- and mid-term training courses for management staff from large and medium-size enterprises on a regular basis. From this group a certain number of promising young and middle-aged managers will be selected to study for their master's degrees at designated colleges of business administration. This unprecedented policy measure shows how much importance is attached to the training of management skills in China.

Training Provided by Academic Institutions

Co-commissioned by industry and enterprises, institutions of higher learning conduct training based on industry's requirements. Generally speaking, university course work in systems engineering, management engineering, and applied science satisfy the needs of management staff. In 1990 alone, 62,000 people from industry were trained in commissioned university programs, thus improving the qualifications of management staff.

In-House Training by Industry

One of the drawbacks common in Chinese industry is that the workers and staff are poorly educated technically and generally, and there is a very low percentage of engineers and other technical people among the employees. According to a survey conducted recently by the All China Federation of Trade Unions, over 40 percent of the workers and staff have only junior middle school education or less, and only 13.1 percent have vocational technical training. Among the scientific and technical personnel working in industry, 4.2 percent have senior professional titles. This has hindered technological progress, product development, and industrial expansion, so there is a pressing need for training.

Various formats are applied by the industrial sector to provide training for workers and staff. Most relied on for training are in-house technology development institutes and technical education institutes. In 1990, technology development institutes attached to large and medium-size enterprises numbered 8,116, and colleges for workers and staff totaled 835. In 1990 alone, 76,639 people completed training at these institutions.

By the end of 1992, a total of 84,000 enterprises with foreign investment were licensed in China, including joint ventures, cooperative production enterprises, and exclusively foreign-owned enterprises. Such enterprises are spread all over the country, with a relatively large number located in the coastal cities. In 1992 alone, direct foreign investment in China reached U.S.$11.16 billion. The joint ventures not only help to introduce new technologies and processes for new product development, but they also introduce new management skills, thus helping to increase China's capability both in technology development and in management. Shanghai Institute of Technology Physics has some positive experience in this respect. The institute set up a joint venture to produce sensors with a Japanese company, Tine Ceramics. A complete set of advanced equipment, appropriate technology processes, and a management system compatible with international norms were introduced. Several hundred thousand sets of Chinese-made sensors found their way into the world market and captured 35 percent of the market within the first year of operation for the joint venture. This shows that joint ventures advance technology development, help market products, and improve management practices.

Radio and television broadcasts from distance universities also play a significant role in technical and management training for industrial enterprises. There are forty such universities in China, and the number of graduates in 1990 totaled 115,896.

In addition, the Chinese Science and Technology Association, a national nongovernmental organization of professionals in S&T, has 364 professional societies that reach across different sectors and industries. The association is also involved in the technical training of workers and management staff. The China Association for International

Science and Technology Cooperation, a newly established nongovernmental organization, is involved in MOT training for technical and professional people both at home and abroad.

Mechanisms for Interaction

Based on the present economic system in China, the government plays the major role in MOT policy making. The development of enterprises, R&D institutions, and universities and colleges, to a certain degree, hinges upon the policies and guidelines adopted by government. For example, the joint development program undertaken by government, industry, and institutions of higher learning is sponsored and presided over by the Office for Trade and Economic Affairs under the State Council, the State Education Commission, and the Chinese Academy of Sciences. National programs such as the Spark Program, the Torch Program, and the High-Tech Development Program, which are sponsored by the State Science and Technology Commission, involve the participation and interaction of industry, institutions of higher learning, and R&D institutions. There are over 2,000 science-policy study institutions across the country providing consulting services to industry, government, and universities and colleges on policy matters.

Over the recent years, in an attempt to create new ways to integrate scientific research, technology development, and industrialization, different market mechanisms have been introduced. The S&T industrial park is an important mechanism, developed to create a favorable environment for business growth.

By the end of 1992, S&T industrial parks numbered 120, of which 52 were approved by the State Council. These parks, accommodating several thousand new and high-tech enterprises, integrate technology development, industrial production, and trade. These industrial enterprises belong to the Chinese Academy of Sciences, to R&D institutions attached to mission ministries and commissions, and to universities and local government. The prominent feature of these industrial parks is that their integration of technology development, industrial production, and trade greatly strengthens the links among industry, universities, and R&D institutions, thus accelerating the commercialization of technological products.

The science parks receive strong government support. Apart from the government's overall planning, policy support, and direct investment, it was the government-sponsored Torch Program that laid the foundation for the development of the parks. Local S&T committees, at the provincial and municipal levels, are executive agencies chiefly responsible for building the parks. According to incomplete statistics, over 12.4 percent of the total income of the parks came from government through project funding, and of all the newly developed products, 17.3 percent were the result of government-funded projects. Recent statistics show that there were 1,268 projects funded by the Torch Program, totaling an investment of RMB Y6.9 billion.

The cooperation between R&D institutions and industrial enterprises plays a vital role in the rapid expansion of industrial parks, which rely on universities and research institutions for trained human resources and R&D. Because of this, all the science parks are located where there is a high concentration of universities and research institutions. Of all the enterprises accommodated in the science parks, 30 percent are businesses operated by universities and research institutions.

In China it is the government that plays the predominant role in policy making and in resource mobilization, and government has the authority to mobilize universities to do the required training. Moreover, the management colleges and adult training centers are sponsored by various ministries to conduct customized training.

The Prevailing Influence on MOT

Because of the fact that China started its industrialization process relatively late, the theories and methodologies related to MOT, as well as the entire technology management system, were in the main introduced from abroad. Due to historical reasons, in the early 1950s, when the industrial and S&T system was to be built, China basically adopted the whole system from the Soviet Union.

Along with the import of 156 major engineering projects from the Soviet Union, there were about a thousand Soviet experts and specialists working in China, and tens of thousands of Chinese students and scholars studying in the Soviet Union and other Eastern European countries. Not only the Soviet industrial system but also its technology management system and related concepts were introduced into China and remained dominant until the late 1970s, when the policy of reform and opening up to the outside world began.

China then opened its doors to North America, Japan, and developed countries in Western Europe, imported their advanced technologies and equipment, and in various ways conducted S&T interaction and exchanges of people. As a result of these activities, a large number of advanced technologies, new concepts, and new management skills came into China, which greatly enriched the thinking of the Chinese people, bringing about substantial advancement in technology management. In the ten years from 1978 to 1988, China imported over one thousand new technologies and tens of thousands of complete sets of equipment valued at U.S.$27 billion. These imported technologies and this equipment played a significant role in economic development and enhanced China's capability in the manufacturing of complete equipment and in technology updating, as well as in R&D in such fields as energy, transportation, communication, and raw materials. Through collaboration with foreign companies on over forty imported engineering projects, China acquired advanced design methodology and specifications and improved its capability in design and production processes and its competence in management.

China has established S&T exchanges with 132 countries in the world and signed intergovernmental agreements of S&T cooperation or economic and trade agreements with 74 countries, including all the developed countries. At the same time, exchanges with nongovernmental organizations and international agencies have been increasing. In 1991 alone, over 14,000 cooperative and exchange projects of an intergovernmental and nongovernmental nature were undertaken; these projects involved 53,000 people. The forms of cooperation and exchange included joint R&D, joint operation of research centers, joint sponsorship of seminars, cooperative research on science policy, joint surveys, and jointly run training centers. All these projects were designed to help the Chinese acquire new management theories, concepts, methodologies, and skills and to build up China's own system by integrating foreign experience with Chinese conditions. Such systems and mechanisms as S&T laws and regulations, the patent system, the nuclear safety system, the S&T statistical indicator system, the science foundation system, the venture capital system, science and industrial parks, business incubators, and the feasibility-study approach are all examples of new concepts and experiences adopted by China from other countries.

China's policy is to take the best concepts and ideas from each country and to invest them with uniquely Chinese features. However, anecdotal evidence shows that the American experience has been a stronger influence on the management of R&D institutions and universities as a result of the many Chinese professionals who were trained in the United States, and of the fact that over 70 percent of the books on technology management in China's libraries were written by Americans. At the enterprise level, the Japanese experience has been more influential because most of the technology and equipment was imported from Japan, and more Japanese commodities are marketed in China. In addition, the geographical proximity of Japan makes it easier for China to introduce Japanese experiences and concepts, especially in the early stages of reform and of the policy to open up China. The influence of Hong Kong and Taiwan has been limited to those regions where their investments were made (Guangdong for Hong Kong, Fujian for Taiwan).

People are the carriers of knowledge, and qualified professionals are essential to the development and mastering of advanced technologies, new concepts, and management skills. China attached great importance to the training of people in its efforts to advance its modernization program. By the end of 1992, a total of 190,000 students and scholars had been sent abroad to acquire new knowledge in S&T and management skills, and about 60,000 of them had returned upon completion of their studies with advanced technologies and new management theories and skills, which greatly improved their qualifications for professional and management positions. In addition, China also devoted much effort to establishing joint training centers with foreign partners abroad to meet the growing need for improved S&T management skills. About 2,500 people have completed a training course in industrial management at the Dalian Industrial Training Center jointly operated by China and the United States since 1981; 650 people have finished training at the Tianjin Enterprise Management Training Center jointly run by China and Japan; 1,800 people have concluded technical training at the Sino-Japan Post and Telecommunication Training Center; and more than 3,000 people have attended training courses at eight Energy Technology Training Centers jointly organized by China and the European Economic Community.

In addition to the above-mentioned forms of training, under the country's new, more open policy, government and nongovernmental organizations have been active in sponsoring short-term training courses in China and abroad to satisfy the demand for trained professionals. Short-term training courses for enterprise managers, S&T management staff, and professionals in statistics and planning have been completed and have proved quite successful.

Training is an important component in national programs such as the Spark Program, which is aimed at the development and application of technology appropriate to boost the rural economy. Encouraged by this program, township and village enterprises are springing up like bamboo shoots after a spring rain. The production value of such enterprises makes up more than one-third of the total industrial value. In such a boom, many farmers who had no formal training at all became enterprise managers. Under the Spark Program, about one million farmers are trained each year for technical and management positions; a considerable number of them become enterprise managers. The training priority of the Torch Program is to prepare professionals for export-oriented industries, with courses in accounting, finance, investment, and trade. This type of training is mainly conducted abroad.

To sum up, the Chinese approach has been to learn advanced management skills and technology from abroad, to apply them to China's own situation, and to create a unique MOT system.

Policy Implications

As China has drawn from the experiences of other countries and integrated management skills developed abroad into Chinese conditions and situations, Chinese MOT has taken shape, and it will be improved and refined by further experience and practice. With the development of a market economy and the transformation of the operation mechanism of state-owned enterprises, especially the growth of the export-oriented economy, technology and production will become more closely linked, and more industrial groups that integrate technology development, industrial production, trade, and finance will be established, as will an increasing number of technology-based firms with equity share. In this rapidly developing situation, MOT in China is facing new challenges—a challenge to adapt the management system to the changes and a challenge to produce a large number of professionals with technological expertise as well as with administrative and management skills.

11
A Comparative View of Pacific Rim MOT Approaches: Implications for Policy

Wong Poh-Kam and Karen Minden

In the previous chapters, nine Pacific Rim case studies were presented, each focusing on a particular economy's conception of the key management of technology (MOT) tasks that it faces, the economic development and public institutional context within which MOT is organized, and the nature of MOT training and other skill-formation mechanisms provided by government, the academic institutions, and the private sector.

Comparing these nine case studies, we come to realize that there are divergent views about the very conception of MOT as a field of research and a set of management practices. This divergence invariably reflects not only that sources of influence and channels of diffusion of MOT theories and practices are different from those in advanced countries but, more important, the different contexts—economic, social, and cultural—within which technology management practices and human resource development occur in these nine economies. In comparing the experiences of the nine economies, therefore, we inevitably learn more about the field of MOT, and the importance of local factors that shape its development in a variety of economic, political, social, and cultural environments.

MOT as an Emerging Field of Study

The theories and practices that comprise the body of MOT knowledge are a collection of concepts used to manage the creation and deployment of technological resources. Even in the advanced industrial countries, it is only in very recent years that MOT has become recognized as an emerging and distinct field of research and education. This is due to the fact that, as a field of scientific inquiry, MOT is an *applied* social science, in the sense that it does not involve the use of a distinct research methodology but rather borrows from existing social sciences (including economics, sociology, and psychology) and extends their applications to the domain of technology. Even as an applied social science, however, MOT cannot be easily separated from other traditional applied fields of management and public policy studies. Since the creation and deployment of technology cuts across traditional functional managerial tasks, MOT differs from traditional business administration subjects such as marketing, personnel, finance, operations, and so on. Similarly, common technology issues cut across traditional public administration jurisdictions (e.g., ministries of education, public infrastructures, industry, and trade) and public policy concerns. In a sense, the emergence of MOT as a distinct field of study in recent years is the result of a radical *paradigm shift* on the part of business and government: the recognition that technology is a resource critical to the competitiveness of firms and nations, and hence of the need to understand and manage technology in an integrative manner or on a "holistic" basis, rather than in traditional compartmentalized ways.

Related to this paradigm shift is the recognition that the traditional notion that science and technology management is the domain of scientists and technologists alone is no longer valid. Excellence in science and technology does not necessarily translate into knowledge and competence in *managing* science and technology, or in *formulating S&T public policies* that are effective. The central concern of MOT is thus to educate and train scientists and engineers to better manage science and technology. A second concern is that traditional managers, who are not necessarily trained in science and technology, better understand science and technology, as well as the scientists and engineers, so that they can manage these resources more effectively.

As in all cases involving paradigm shift, there is resistance to this shift among practitioners of existing disciplines. This is one factor that has affected the slow growth of MOT as a recognized field of research and education even in the advanced industrial countries. Another factor is the lack of a sufficient number of social scientists and management researchers with some training in the hard sciences and engineering studies that underlie modern technologies. This lack of familiarity, besides limiting the ability of scientists and researchers to pursue research that requires domain-specific knowledge and insight, also creates "outsider" credibility problems among potential users of research results. Other retarding factors include the traditional compartmentalization of technology into distinct professional disciplines (e.g., civil engineering, electrical engineering, mechanical engineering), which tends to prevent the emergence of a common conception of the generic tasks of technology management among technologists of different professional disciplines; the lack of a sufficient base of accumulated empirical research attesting to the efficacy of MOT theories in the real world; and the relatively small number of researchers and educators who can serve as agents in transmitting MOT knowledge. From the perspective of the history of science, these problems are all symptomatic of emerging paradigms.

Diffusion of MOT in the Pacific Rim

Given the nascent stage of MOT as a field of research and education even in the advanced industrial nations, we can expect a priori that MOT as a discipline would be even less developed in the less advanced economies. It is thus not surprising that our survey of the comparative experiences of Canada and the newly industrializing economies in the Pacific Rim shows a wide divergence in the conception of MOT itself, as well as in the degree of sophistication with which it is practiced by government, academic institutions, and business firms.

In his classic study, *The Diffusion of Innovation*, Everett Rogers concludes that despite Americans' favorable disposition toward science and technology, there is generally a considerable lag between the time an innovation is made known and the time it reaches wide acceptance, in spite of demonstrable economic benefits (Rogers 1962, 2). Similarly, we have seen in this collection of nine case studies that while there is a general awareness of the desirability of taking advantage of MOT theories and practices from the advanced industrial countries, the pace and pattern of adoption of such theories and practices are quite uneven among the nine Pacific Rim economies in this study. As explained earlier, MOT is not a mature body of knowledge, but rather a recent paradigmatic conception that is still evolving, thus hindering its broad diffusion and wide acceptance. Furthermore, the adoption of new MOT practices often requires adaption to conditions specific to particular socioeconomic and cultural contexts, as shown in the nine case studies. The factors affecting the development of technology management are highly complex. A great variety of cross-cultural issues affect the transfer of MOT practice, and there is great diversity in the pattern of adaptation and the degree of indigenization of MOT by the three sectors—academe, industry, and government—that are concerned with MOT in each economy.

Developing Technology Managers

The ultimate manifestation of the diffusion of MOT as a body of knowledge is the embodiment of MOT skills among practicing technology managers in both the public and private sectors. A key indicator of the success of MOT is therefore the area of human resource development—the extent to which technology managers are trained in specific MOT skills that allow them to perform more effectively.

Nine comparative case studies make it amply clear that the specific mix of technology management tasks varies among economies at different stages of development, and consequently, so does the specific mix of MOT skills required. Moreover, as these economies continue to develop technologically, a key challenge they will face is how to adapt their MOT skill formation mechanisms in response to the changing mix and sophistication of technology management tasks that need to be performed.

Diversity of MOT Development Concerns in the Pacific Rim

Disaggregated statistics on the number of technology managers and the mix of MOT tasks that they perform are not available for the nine economies covered in the study. Nonetheless, some available comparative S&T human resource development indicators as given in Table 11.1 are suggestive of the different MOT challenges facing the nine economies. As can be seen, Canada, as an industrialized North American economy, is the most advanced among the nine cases covered in this book in terms of the percentage of professional, technical, and managerial occupations in the workforce, the percentage of scientists, engineers, and technicians in the total labor force, the proportion of university graduates in science and engineering, and the ratio of research and development (R&D) spending to gross domestic product (GDP). The four newly industrialized economies (NIEs), Hong Kong, Taiwan, Korea, and Singapore, rank next. Although differences exist among the NIEs, none dominates all others on all indicators. The three ASEAN countries, Malaysia, Indonesia, and Thailand, although developing rapidly in recent years, are still clearly behind the four NIEs in terms of all the indicators given in Table 11.1. China appears to rank somewhere between the averages of the Asian NIEs and the ASEAN countries, although given its sheer size, it should really be regarded as being in a different league.

The above broad differences in S&T development are clearly reflected in the differing concerns with MOT challenges in the nine case studies. Given the relatively high level of development of the Canadian economy, it is not surprising that Canada appears to be most concerned with MOT as a factor in improving national competitiveness. While some of the less advanced economies are still preoccupied with developing enough scientists, engineers, and technicians, Canada is more concerned with how to manage the considerable S&T human resources it already has. As such, the chapter on Canada focuses prominently on MOT research and education in the universities.

Although still concerned with developing S&T human resources in general, the NIEs are all increasing their efforts to improve MOT. As in other spheres of economic development, a diversity of patterns can be discerned among the NIEs with respect to MOT skills development. Typical of its laissez-faire policy, the government of Hong Kong does not have a history of active intervention in S&T matters beyond providing educational infrastructure, and hence there has hitherto been little concern with developing technology managers in the public sector, while MOT skills development in the private sector has been left primarily to the universities and the private sector itself. In contrast, there is significant public sector involvement in S&T policy and management in both Korea and Taiwan, with Singapore somewhere in between: while trained experienced foreign technology managers have been heavily used to meet the MOT requirements of the advanced S&T sectors of the economy, including the government research institutes, there has been increasing emphasis on expanding MOT training in the local universities, as well as within the large corporations.

The chapters on the ASEAN countries indicate that these countries

Table 11.1

Comparative S&T and Managerial Workforce Indicators

Country	(1) Percentage Professional, Technical, and Administrative/ Managerial Occupations in Total Employment	(2) Percentage of Scientists, Engineers, and Technicians in Labor Force (1986–90)	(3) Ratio of Science and Engineering Degrees to Total Bachelor's Degrees (1988–90)	(4) (R&D/GDP) Ratio
Canada	27.7 (1992)	17.7	27	1.50 (1992)
Singapore	29.5 (1992)	2.4	53	1.27 (1992)
Hong Kong	14.0 (1991)	4.1	39	n.a.
Taiwan	23.9 (1992)	2.6	47	1.70 (1991)
Korea	25.9 (1990)	4.7	29	2.02 (1991)
Thailand	5.4 (1991)	0.1	18	0.17 (1992)
Malaysia	12.0 (1992)	n.a.	28	0.80 (1992)
Indonesia	4.2 (1991)	0.1	11	0.25 (1993)
China	7.1 (1991)	0.9	43	0.71 (1992)

Sources: (1), (2), (3)—Pacific Economic Cooperation Council (PECC), Human Resource Development Outlook, 1994–95; (4)—PECC, S&T Profiles, 1994.

are still preoccupied with increasing the S&T human resources in general, with a weaker focus on MOT skills development specifically. In Malaysia and Thailand, for example, the general shortage of skilled engineers and technicians needed to service rapidly growing manufacturing industries has preoccupied the government. While the importance of MOT skills has been recognized, it has not been translated into any significant policy, and university initiatives have been slow. However, as governments continue to become more involved in S&T planning, policy intervention, and public R&D programs, bottlenecks in MOT skills will become more keenly felt. Malaysia and Indonesia, in particular, will need to move beyond reliance on foreign consultants in public sector MOT.

Finally, China faces the most acute challenge in terms of developing management expertise to harness the considerable scientific and technical human resources it has built up through a traditionally strong emphasis on S&T education. The transition from a command economy to a market economy requires a very dramatic change in the orientation and methods of management, while years of inward-looking economic policy deprived China of access to experienced technology managers from abroad and exposure to cutting-edge MOT practices. Recently China has focused on harnessing these resources and adapting them to its own development needs.

The Role of the Public, Private, and Academic Sectors

The stage of economic development is clearly an important variable affecting the overall level and sophistication of MOT skills demanded in each of the economies, and consequently, the level of overall awareness of MOT, and the degree of emphasis given to MOT skills formation in general. In addition, the case studies suggest that the overall pace of diffusion and development of MOT in the Pacific Rim is strongly influenced by the public sector, the private sector, and local tertiary educational institutions. Each of the chapters in this book has sought to identify the specific roles that these three sectors have played. The diversity of these roles and their relative importance in the different economies preclude systematic generalizations, but the following broad observations can be made.

First, in terms of the role of the public sector, an important factor appears to be the level and sophistication of government understanding of MOT as a source of competitive advantage, and how this understanding is effectively translated into policies and programs targeted at enhancing MOT skills formation in the private sector. In this regard, the NIE governments (with the possible exception of Hong Kong) generally have a more sophisticated awareness of the importance of MOT and hence pursue more systematic policies to promote MOT skills formation (e.g., policies to recruit experienced technology managers from abroad for top posts in government research institutes; sending government S&T planners abroad to study MOT courses; government training and seminar programs to educate managers from business and government research institutes in MOT skills). In contrast, ASEAN governments have not moved much beyond policy intention announcements and tend to be hampered by a lack of policy coordination (particularly in the case of Thailand) or constrained by indigenization policies that limit the use of foreign talent. Despite a high awareness of MOT on the part of the Canadian government, the chapter on Canada raises the issue of the effectiveness of government in enhancing MOT skills formation in the private sector.

Besides government responses, the degree of responsiveness of the higher-education systems appears to vary across the nine Pacific Rim economies. In several of the cases covered in this book, public tertiary institutions appear to lag behind the needs of the market due to their weak links with private industry, and they are relatively slow in innovating new MOT training programs due to the inertia of traditional compartmentalization. Sheltered from competition, some of the local tertiary institutions appear to exhibit stronger resistance to the MOT paradigm shift than local firms in the private sector. The problem thus appears not to be purely one of financial constraints—countries like Malaysia, for example, devote substantial financial resources to university education—but relates to how effectively the academic institutions themselves allocate and use these resources.

In the private sector, while most multinational corporations (MNCs) have by and large been able to take care of their own MOT training needs in-house, the contributions have been much weaker among the smaller local enterprises. Even in Canada, the most advanced country among the nine economies covered in this book, the diffusion of MOT ideas and practices appears to be rather slow among many small and medium-size enterprises. Their unwillingness to invest in MOT skills formation has weakened the viability of a number of government programs designed to promote MOT training. Although this issue is not systematically examined in the other chapters, the problem is likely to be more severe in most of the other economies.

Future MOT Challenges

As pointed out earlier, the level of technological development in an economy influences the nature of the MOT challenge that economy faces. The predominant need in Indonesia, Malaysia, and Thailand is to promote technology transfer from MNCs, to upgrade the manufacturing process technology capability of local small and medium-size enterprises, to expand the government's research institutions and begin building an indigenous R&D base, and to improve the planning and provision of S&T infrastructure to assist the private sector in technological upgrading. On the other hand, industries in Taiwan, Singapore, Korea, and Hong Kong are moving increasingly into higher value-added activities and requires technological advances that will allow them to move increasingly into new product technology innovation and commercialization. Formerly preoccupied with technology acquisition from more industrially advanced nations, the NIEs are now moving toward further strengthening their own indigenous research and development base, expanding their technology capability through strategic alliances with firms from advanced industrial countries, and transferring manufacturing operations and some of their own technologies to less advanced countries. The supporting public S&T infrastructure has to be correspondingly expanded in order to provide assistance in such areas as technology export, intellectual property rights protection, international R&D collaboration, and participation in international standards setting.

The underlying theme of this variety of challenges is that MOT strategies must be adapted to the specific needs of an economy at a particular stage of development. Just as the concept of "appropriate technology" has evolved in the realm of technology acquisition, so the notion of "appropriate MOT" must be applied to the different levels of technological capability outlined in chapter 1.

Despite the diversity of development contexts and constraints, one common issue that appears in each of the case studies is the challenge to integrate technical skills and management skills. This need occurs in academe, industry, and government ministries involved in S&T development and promotion. It is a matter of bridging not only national cultures but the cultures of professional specialties as well.

Policy Implications

In each of the nine chapters, a wide range of policy issues was raised to address each economy's specific concerns. The diversity of the issues raised reflects the variety of economic development contexts represented, a variety which precludes simplistic generalizations applicable to all economies. Nonetheless, there appear to be a number of generic policy issues that deserve the attention of policy makers in universities, government, industry, and international organizations.

University Sector

While tertiary education institutions in the Pacific Rim vary in their responsiveness to the demand for MOT training, they have generally been slow to innovate new programs and to integrate technological and management education, and to forge adequate links with private industry. There is thus an urgent need for university administrators to promote cooperation between management and engineering faculties in terms of research and training programs and to foster closer university-industry collaboration. It is likely that to achieve these goals, significant changes will need to be made in university policy; the incentive and reward system may have to be altered, as may staff recruitment criteria and policies and research-funding priorities.

While universities in Pacific Rim developing economies have relied upon links with universities in the advanced industrial countries, particularly the United States and Europe, to upgrade their teaching and research capabilities, there have been few such links among these economies themselves. In view of their similar sociocultural contexts and levels of economic development, there should be significant scope for these East Asian economies to share best MOT practices and resources. In this regard, a major policy implication is that closer academic exchange and collaboration should be fostered among the universities in the East Asia region itself. In particular, the exchange of teaching materials "customized" for the East Asian region should be actively encouraged.

Government Sector

While the extent to which government provides S&T infrastructural support for private industry varies in Pacific Rim economies, the inevitable trend is for government involvement to increase. As government S&T tasks become increasingly complex, it is important that technology managers in the government become better trained in MOT skills. Technology managers in public research institutes also need to be made more aware of the needs of industry and to be more responsive to the demands of competition. Singapore's success in attracting experienced technology managers from overseas suggests that governments should consider adjusting their immigration and public-sector-employment policies to allow recruitment of expertise from an international pool. Other relevant policy issues include closer collaboration among government, universities, and industry in S&T development, and policies to encourage universities to be more responsive to MOT training needs.

Private Sector

In terms of promoting MOT in the private sector, the case studies show clearly that it is local small and medium-size private enterprises that are most in need of strengthening their MOT capabilities, yet, paradoxically, they are typically the most reluctant to invest in the development of MOT skills. Part of the problem is that many existing MOT training programs do not specifically address the needs of SMEs. A more serious problem is that many entrepreneurs have a strong need to exert control over their own enterprises and do not believe in the value of professional management.

A conscious attempt to use local success models to promote greater diffusion of MOT awareness and skills has been a prominent feature of government policy in Korea, Taiwan, and, more recently, Singapore. Publicizing success stories that illustrate how better MOT practice has contributed to the growth of local high-technology industries is one approach that may usefully be studied for adaptation in other developing economies.

International Agencies

International development agencies such as the World Bank Economic Development Institute, the Asian Development Bank, and the United

Nations Industrial Development Organization (UNIDO) have traditionally focused on general economic management, financial management, and industrial development, with little emphasis on MOT human resource development. These agencies can help to promote the importance of technology managers in the industrial development process. They could focus in particular on fostering an international pool of experts to develop training materials and programs for small and medium-size enterprises, which do not have the capacity to develop their own expertise. They can also help to disseminate training materials based on the MOT practices that have been successful in the NIEs, which are likely to be more relevant in other developing economies than those derived from the experiences of high-tech companies in advanced countries.

Another policy implication for international agencies is that regional collaboration on MOT research and the best MOT practices, especially among Pacific Rim economies at similar stages of economic development, should receive strong support.

Concluding Remarks

Managing technology is a complex endeavor. As in the case of all management approaches, the basic theoretical principles and techniques of MOT may be universal, but the most useful application of these techniques, the preconditions for their successful adoption, the adaptation of the principles to specific practice, and the empirical validation of these practices are dependent on the specific context of the firm or whole economy. A well-known example is the case of total quality management (TQM), a concept developed in the West but successfully adopted in Japan and widely employed as a management practice. In some cases, formulation of generalized principles and concepts follow, rather than precede, their actual practice. For example, the concept of time-based competition was first extracted from studying the practice of a number of leading Japanese firms. The emergence of these and other new MOT practices from industry leaders outside the United States and Europe allows us to better understand the importance of the holistic context within which best practices of MOT can evolve. Progress in MOT research therefore comes not only from developing better concepts and techniques, but also from a deeper understanding of the contexts in which MOT principles are to be applied and adapted. In this regard, one principal contribution of this collection of Pacific Rim case studies is to help MOT educators and researchers to better understand the diversity of contexts in the region.

The manner in which each of the economies in this study responds to the challenge to develop the resources to manage technology will depend on a variety of factors: level of development, foreign influence, major economic and political constraints, the role of multinational companies, existing educational infrastructure, policy predispositions, and the needs and responses of the private sector. The preceding chapters illustrate that the variables are complex and differ widely from case to case. However, this comparison of approaches used to develop technology managers has served a dual purpose: it has sharpened our focus on our respective technology systems, and it has heightened our awareness of the benefits of and possibilities for regional cooperation for the development of the skills required by the demanding pace of technological advancement.

Reference

Rogers, Everett M. 1962. *Diffusion of Innovations*. New York: The Free Press.

Appendix: List of Participants at Conferences on Developing Technology Managers in the Pacific Rim

Meeting Held at the National University of Singapore, November 26, 1992

Dr. George Abonyi, Senior Advisor, Canadian High Commission, Singapore

Mr. Arun Abraham, Science and Technology Management, Regional Program Officer, International Development Research Centre, Southeast Asia Regional Office

Dr. Charles H. Davis, Principal Program Officer, Program on Innovation Systems Management, International Development Research Centre, Ottawa

Mr. Sree Kumar, Senior Fellow, Institute for Southeast Asian Studies, Singapore

Dr. Karen Minden, Visiting Fellow, Centre for Management of Technology, National University of Singapore, and Institute for Southeast Asian Studies, Singapore

Dr. Ng Chee Yuen, Senior Fellow, Institute for Southeast Asian Studies, Singapore

Professor Pang Eng Fong, Faculty of Business Administration, National University of Singapore

Dr. Wong Poh-Kam, Director, Centre for Management of Technology, National University of Singapore, Singapore

Dr. Zhao Hongxin, Lecturer, Faculty of Business Administration, National University of Singapore

Meeting Held in Singapore, March 11–12, 1993

Mr. Arun Abraham, Regional Program Officer, Science and Technology Management, International Development Research Centre, Southeast and East Asia Regional Office, Singapore

Dr. Chatri Sripaipan, Director, Science and Technology Development Program, Thailand Development Research Institute, Bangkok, Thailand

Dr. Yegin Chen, Department of Operations and Systems Management, Faculty of Business Administration, Chinese University of Hong Kong, Hong Kong

Professor Jong-Tsong Chiang, Department of Business Administration, National Taiwan University, Taipei, Taiwan

Dr. Charles H. Davis, Principal Program Officer, Science and Technology Policy and Management, International Development Research Centre, Ottawa

Mme. Dong Guilan, Associate Research Professor, and Director, Office of Academic Affairs, National Research Centre for Science and Technology for Development, State Science and Technology Commission, Beijing, China

Mr. Jim Feir, Counsellor, Commercial Division, Canadian High Commission, Singapore

Profesor Foo Say Wei, Department of Electrical Engineering, Faculty of Engineering, National University of Singapore, Singapore

LTC (Res) Benny Goh, Director, Planning and Policy, National Science and Technology Board, Singapore

Dr. Jingjai Hanchanlash, Director, International Development Research Centre, Southeast and East Asia Regional Office, Singapore

Dr. Ji-Soo Kim, Department of Management and Policy, Korea Advanced Institute of Science and Technology, Taejon, Korea

Mr. Sree Kumar, Senior Fellow, Institute for Southeast Asian Studies, Singapore

Dr. Cornelis Luhulima, Senior Researcher, Indonesian Institute of Sciences (LIPI), Jakarta, Indonesia

Ms. Michelle McCormack, Third Secretary, Commercial Division, Canadian High Commission, Singapore

Ms. Susan J. MacDonald, Director, Office of the Agent of Ontario, Singapore

Dr. Karen Minden, Director of Policy Studies, Centre for Higher Education Research and Development, Visiting Fellow, Centre for Management of Technology, and Institute for Southeast Asian Studies, Singapore

Mr. Nguyen Thanh Ha, Research Fellow, Institute for Science Management, Ministry of Science, Technology and Environment, Hanoi, Vietnam

Ms. Rahimah Said, Deputy Director, Science and Technology, Ministry of Science, Technology and Environment, Kuala Lumpur, Malaysia

Dr. Chan Heng Chee, Director, Institute for Southeast Asian Studies, Singapore

Dr. Manjeet Singh, Assistant Director, Industry and Technology Relations Office, National University of Singapore, Singapore

Mr. Richard Smith, Associate Director, Centre for Policy Research on Science and Technology, Simon Fraser University at Harbour Centre, Vancouver

Professor Soon Teck Wong, Department of Economics and Statistics, Faculty of Arts and Social Sciences, National University of Singapore, Singapore

Dr. Vu Cao Dam, Director, Institute for Science Management, Ministry of Science, Technology and Environment, Hanoi, Vietnam

Dr. Wong Poh-Kam, Director, Centre for Management of Technology, Faculty of Business Administration, National University of Singapore, Singapore

Dr. Yap Chee Meng, Department of Industrial and Systems Engineering, Faculty of Engineering, National University of Singapore, Singapore

Dr. Zhao Hongxin, Lecturer, Department of Business Policy, Faculty of Business Administration, National University of Singapore, Singapore

Index

B

C

O

P